# The Actual and the Possible

## MIND ASSOCIATION OCCASIONAL SERIES

This series consists of carefully selected volumes of significant original papers on predefined themes, normally growing out of a conference supported by a Mind Association Major Conference Grant. The Association nominates an editor or editors for each collection, and may cooperate with other bodies in promoting conferences or other scholarly activities in connection with the preparation of particular volumes.

*Director, Mind Association*: Julian Dodd
*Publications Officer*: Sarah Sawyer

RECENTLY PUBLISHED IN THE SERIES:

*Thinking about the Emotions*
Edited by Alix Cohen and Robert Stern

*Art, Mind, and Narrative*
Edited by Julian Dodd

*The Social and Political Philosophy of Mary Wollstonecraft*
Edited by Sandrine Bergès and Alan Coffee

*The Epistemic Life of Groups*
Edited by Michael S. Brady and Miranda Fricker

*Reality Making*
Edited by Mark Jago

*The Metaphysics of Relations*
Edited by Anna Marmodoro and David Yates

*Thomas Reid on Mind, Knowledge, and Value*
Rebecca Copenhaver, Todd Buras

*The Highest Good in Aristotle and Kant*
Joachim Aufderheide, Ralf M. Bader

*Foundations of Logical Consequence*
Edited by Colin R. Caret and Ole T. Hjortland

*The Highest Good in Aristotle and Kant*
Edited by Joachim Aufderheide and Ralf M. Bader

# The Actual and the Possible

*Modality and Metaphysics in Modern Philosophy*

EDITED BY
Mark Sinclair

OXFORD
UNIVERSITY PRESS

OXFORD
UNIVERSITY PRESS

Great Clarendon Street, Oxford, OX2 6DP,
United Kingdom

Oxford University Press is a department of the University of Oxford.
It furthers the University's objective of excellence in research, scholarship,
and education by publishing worldwide. Oxford is a registered trade mark of
Oxford University Press in the UK and in certain other countries



First Edition published in 2017



Published in the United States of America by Oxford University Press
198 Madison Avenue, New York, NY 10016, United States of America

British Library Cataloguing in Publication Data
Data available

Library of Congress Control Number: 2017944692

ISBN 978–0–19–878643–6

# Contents

# Acknowledgements

My thanks are due to James Clarke who organized with me the 2013 British Society for the History of Philosophy annual conference at the University of York, and to the then President of the Society, Pauline Phemister, who offered insightful guidance throughout. My thanks are also due to the Mind Association for the award of a Major Conference Grant.

*MWS*

# Note on the Contributors

THOMAS BALDWIN is Emeritus Professor of Philosophy at the University of York and former editor of *Mind.* His recent work includes *G. E. Moore: Early Philosophical Writings*, 'G. E. Moore and the Cambridge School of Analysis' and 'Truth: British Idealism and its Analytic Critics'.

JOHN DIVERS, Professor of Philosophy at the University of Leeds, is the author of various articles on modality (in *Mind, Noûs, Philosophy and Phenomenological Research*, etc.), *Possible Worlds* (Routledge, 2002) and *Necessity After Quine* (contracted to OUP).

MOGENS LÆRKE is Senior Researcher at the CNRS in France, affiliated at the ENS de Lyon. He is the author of *Leibniz lecteur de Spinoza* (Paris, 2008) and *Les Lumières de Leibniz* (Paris, 2015), of more than fifty articles principally on early modern philosophy, and has edited five volumes including *Philosophy and Its History* (Oxford University Press, 2013).

JESSICA LEECH is Lecturer of Philosophy at King's College London. She gained her doctorate from the Universities of Geneva and Sheffield. She works mainly on the philosophy of modality and Kant's theoretical philosophy.

STEPHAN LEUENBERGER is Senior Lecturer in Philosophy at the University of Glasgow. He gained his PhD at Princeton, and received the Oxford Studies in Metaphysics Younger Scholar Prize (2006) and the Lauener Prize (2009). He has published mainly on the topics of modality and supervenience.

OHAD NACHTOMY is Associate Professor and Chair at Bar-Ilan University. He is the author of *Possibility, Agency, and Individuality in Leibniz's Metaphysics* (Springer, 2007); the editor (with Justin Smith) of *The Life Sciences in Early Modern Philosophy* (Oxford University Press, 2014) and *Machines of Nature and Corporeal Substances in Leibniz* (Springer, 2010); and has published some forty articles in the history of early modern philosophy and the philosophy of biology.

PETER SIMONS is Emeritus Professor of Philosophy at Trinity College Dublin and is Honorary Professor at the University of Salzburg. He is the author of two books and over two hundred articles, and specializes in metaphysics, the history of logic, and central European (Polish and Austrian) nineteenth- and twentieth-century philosophy.

MARK SINCLAIR is Reader in Philosophy at the University of Roehampton and Associate Editor at the *British Journal for the History of Philosophy*. He has published on the history of modern French and German philosophy in *Journal of the History of Philosophy*, *History of Philosophy Quarterly*, *Archiv für Geschichte der Philosophie*, and the *Journal of the History of Ideas*.

Christopher Yeomans is Professor of Philosophy and Head of Department at Purdue University. He is the author of *Freedom and Reflection: Hegel and the Logic of Agency* (Oxford University Press, 2012) and *The Expansion of Autonomy: Hegel's Pluralistic Philosophy of Action* (both from Oxford University Press).

# Editor's Introduction

This volume offers a selection of essays on modality and the metaphysics of modality in the history of modern philosophy from the seventeenth to the twentieth centuries. The essays were mainly selected from contributions to the 2013 British Society for the History of Philosophy annual conference bearing the title 'The Actual and the Possible', which had a broad remit concerning modality and the metaphysics of modality throughout the history of philosophy. The narrower historical focus of this volume arose from a preponderance of papers on the history of modern philosophy, which seemed to express a need for a published work that would revisit key moments in the history of modern modal doctrines as well as illuminate lesser-known moments of that history, and this with the aim of contextualising, and perhaps even of offering alternatives to dominant positions within the contemporary philosophy of modality. Hence the volume contains not only new scholarship on the early-modern doctrines of Baruch Spinoza, G. W. F. Leibniz, Christian Wolff and Immanuel Kant, but also work relating to less familiar nineteenth-century thinkers such as Alexius Meinong and Jan Łukasiewicz, together with essays on celebrated nineteenth- and twentieth-century thinkers such as G. W. F. Hegel, Martin Heidegger and Bertrand Russell, whose modal doctrines have not previously garnered much attention. The volume thus covers a variety of traditions, and its historical range extends to the end of the twentieth century, since it addresses the legacy of Willard Van Orman Quine's critique of modality within recent analytic philosophy.

The modal notions are those of possibility and necessity, together with the related notions of impossibility and contingency, and the metaphysics of modality, as commonly understood, concerns the grounds or truth-conditions of statements containing these notions. Such statements pervade both our ordinary talk and our scientific discourse. That there could be more customers in the café than there actually are, that I could get to work in a number of ways, that the laws of nature cannot be broken, that the glass might break if it falls, that unicorns do not actually exist but could exist, are all assertions that seem to meet with common assent. The task for philosophy, however, is to determine the content of such statements, whether they can be true or false like those concerning the actual world, and what it is that could ground their claims to truth. In taking up this task, the philosopher has to address the fact that we speak modally according to a varying scope: travelling faster than the speed of light, for example, may well be metaphysically possible—possible in the widest sense—but it is not physically or nomologically possible, i.e. not compatible with the laws of nature. The philosopher also has to address how modal statements can be *de re* ('John is necessarily human'), where the modality is ascribed to the thing as such, as well as *de dicto* ('It is necessary that 5 + 2 =7'), where the modality attaches to the statement as a whole.

Much of the work on the metaphysics of such truth-concerning or 'alethic' modality in recent decades—work which stands at the forefront of the revival of metaphysics within analytic philosophy—has focused on a paradigm of 'possible worlds' that has a clear historical lineage. For in later medieval philosophy, as a wealth of twentieth-century historical scholarship has shown,[1] a temporal or 'statistical' interpretation of modal terms deriving from ancient philosophy was replaced by a different paradigm that would later be articulated in terms of 'possible worlds'. According to the 'statistical' paradigm, modal terms were interpreted extensionally, with the necessary understood as that which is always actual, the impossible as that which is never actual, the possible as that which is actual at some point in time, and the contingent as that which is actual but which will not always have been actual. From this perspective, according to what Arthur Lovejoy famously named a 'principle of plenitude',[2] what can be, will at one point have been—and if it is not at some point actual, then it is not possible. Under pressure from the idea of a Christian God freely choosing between alternate histories of a created world, however, this temporal grounding of modal notions gave way to a paradigm admitting unrealized possibilities and simultaneous alternate possibilities—and this paradigm received its grandest development in Leibniz's notion of possible worlds arrayed before the understanding of a God, who, by virtue of his goodness, actualizes only the best one among them.

Interest in Leibniz's conception of possibility was rekindled in the mid-twentieth century by Rudolf Carnap's project of a semantics for languages with modal operators by means of 'state descriptions', understood as maximal consistent sets of sentences; these, Carnap suggested, could be thought as akin to Leibniz's 'possible worlds'. Several logicians in the 1950s and early 1960s took up in earnest this Leibnizian approach in developing a model theoretical semantics for modal logic. The most prominent of them was Saul Kripke, who presents a set theoretical model structure for propositional modal logic that can be intuitively understood as comprising the actual world, the set of all possible worlds, and the relative possibilities between these worlds (which would later be named the 'accessibility relation'). On this basis, a proposition can be understood as possible if it is true in at least one of

[1] For a recent expression of this scholarship, see Simo Knuuttila, 'Modality' in *The Oxford Handbook of Medieval Philosophy*, ed. J. Marenbon (Oxford: Oxford University Press, 2012), 312–41.

[2] See Arthur Lovejoy, *The Great Chain of Being* (Cambridge MA: Harvard University Press, 1936) and, for critical responses, Simo Knuuttila (ed.), *Reforging the Great Chain of Being* (Dordrecht: D. Reidel, 1980).

these worlds, and as necessary if it is true in all of them. As well as clarifying our ordinary modal discourse and providing truthmakers—however merely theoretical such truthmakers may have been supposed to be—for modal statements, such a set theoretical structure of 'possible worlds' has the important result of showing how various combinations of the properties of the accessibility relation, namely reflexivity, transitivity, and symmetry, provide semantics validating the axioms of the various systems of modal logic formulated earlier in the twentieth century.

Possible worlds talk is a powerful philosophical tool, and it was not long before one philosopher, David Lewis, interpreted this talk in the most literal fashion. According to his modal realism, if talking donkeys could exist, then they really exist in another world. For every possibility in this world, there exists another world in which that possibility is real; but the possible worlds in which they exist, though they are much like our actual world, are spatially unrelated to it and unobservable. Though certainly strange, such modal realism permits the reduction of a quantified modal language to a first-order non-modal language, to well-understood existential and universal quantification, and thus, for Lewis, in terms of its elegance and explanatory power, it is the best available theory of possibility. Of course, a variety of less incredible surrogates for these really existing possible worlds have been proposed in the literature: not just maximal consistent sets of sentences, but propositions, properties, pictures, and imaginative fictions have all been proposed as constituting, as Lewis puts it, Ersatz worlds.[3] Given, however, the lack of consensus on the merits of any of these options, or even on whether the sort of reductive theory of modality that Lewis proposes is a desirable goal, it is understandable that the very idea that modal discourse has truth conditions has been challenged. At the opposite end of the spectrum to a Lewisian realism, an anti-realist position, according to which modal talk is not in any objective sense truth-apt, and is merely an expressive projection of our attitudes, could well save us from much toil, that however honest, may ultimately be fruitless.[4] At the same time, it is not surprising that modality has recently been dethroned, as it were, in analytic philosophy with the claim that it is not the most basic ontological category, that in what there is modality 'does not go all the way down', according to the argument that the essence of a thing cannot be accounted for in terms of *de re* necessity.[5] It is no more surprising that the contemporary philosophy of powers has sought to account for the content of modal claims by means of a this-worldly, Aristotelian paradigm of potentiality as the ultimate *explanans* in any metaphysics

[3] For an exhaustive survey of these alternatives to Lewis's 'genuine realism', see *Possible Worlds* (London: Routledge, 2002) by John Divers, who very helpfully commented on a draft of this introduction, as did the anonymous reviewers of the manuscript.

[4] See, in particular, Simon Blackburn, 'Morals and Modals' in *Essays in Quasi-Realism* (Oxford: Oxford University Press, 1993), 52–74.

[5] The reference here is to the work of Kit Fine, but see John Diver's analysis of Fine's position in his contribution to this volume.

of modality.[6] With such notions of essence and potentiality, a renewed taste for metaphysics in analytic philosophy has led from Leibniz all the way back to Aristotle.

It is against this background that the present volume attempts to re-assess modal doctrines in the modern period, not only with the aim of further elucidating and situating Leibniz's thinking in the context of Early Modern philosophy, but also with the goal of assessing alternate conceptions of the modal notions in later centuries. The volume focuses on the metaphysical modalities—in contrast to, say, epistemological or deontological modalities—and although its approach is exploratory, open-minded and even eclectic, contributors to the original conference were encouraged to reflect on how modal doctrines might impact on the idea of actuality. Actuality is, after all, according to the philosophical tradition at least (as in Kant, for example), one of the modal categories, and a particular aim of the volume is to address how the inter-definable notions of possibility and necessity might affect how we should understand both actuality and the ontological notions closely related to it, namely existence and being. Chapters 1, 3, 4, 5 and 8 all address this broad issue in one way or another, to a greater or less extent, and this explains the order of the terms in the title of the volume: *The Actual and the Possible: Modality and Metaphysics in Modern Philosophy.*

In the first essay, 'Aspects of Spinoza's Theory of Essence: Formal Essence, Non-Existence, and Two Types of Actuality', Mogens Lærke criticizes what he takes to be a recent wave of 'Platonizing' interpretations of the Dutch philosopher. According to these interpretations, Spinoza belongs to an essentialist tradition holding that the essences of things exist outside of time and beyond the actual world; what he names *essentia formalis,* formal essence, would, on these accounts, be something distinct from, or a distinct way of being, to *essentia actualis*, actual essence. Such Platonizing interpretations have significant consequences for our understanding of possibility and actuality in Spinoza's philosophy: formal essences could thus be understood as populating a realm of *possibilia*, separate from the actual world, and being or existence would therefore have to be said in different ways of *possibilia* and actually existent things. Such consequences are controversial, since Spinoza is otherwise committed to an actualist doctrine that possibilities, as merely a function of our ignorance, in no way exist; and we have good reasons to suppose that he is committed to the univocity of being, according to which if things exist, they exist in one way only, without degrees or gradation. In holding that Spinoza is indeed consistently committed to these doctrines, Lærke argues that Spinoza does not oppose formal essence to actual essence, and that he characterizes essence as actual only to describe the power or striving—*conatus*—constitutive of essence itself. This is to say that Spinoza understands actuality as power or striving. At the same time, Lærke argues,

[6] See, for example, Jonathan D. Jacobs, 'A Powers Theory of Modality; or, How I Learned to Stop Worrying and Reject Possible Worlds', *Philosophical Studies* 151: 227–48, and Barbara Vetter, *Potentiality* (Oxford: Oxford University Press, 2015).

the formal essences of non-existent things, which Spinoza mysteriously claims are eternally contained in the divine substance, should not be understood as non-actual or possible things, but as things that are grasped only qua non-existent through the actually existing 'cause or reason of their non-existence'.

Stephan Leuenberger, in 'Wolff's Close Shave with Fatalism', addresses Christian Wolff's modal thinking, and asks whether it is compatible with his denial of fatalism. Wolff had become a *cause célèbre* throughout Europe after Friedrich Wilhelm I understood him to be advancing a fatalist doctrine undermining personal responsibility and consequently morality, and ordered him to leave Prussia on pain of death. Leuenberger assesses Wolff's modal doctrines in their own terms, independently of Leibniz, and discovers a fundamental tension: Wolff's *Ontology* defines philosophy as the 'science of the possibles insofar as they can be', and he certainly claims that what is possible does not have to exist, but his definition of the impossible as that which contradicts a true sentence commits him to the doctrine that every falsity is impossible, and thus to the idea that everything, as it now is, is necessary. Wolff seems to have been aware of this, but he also seems to have thought it possible to avoid fatalism by distinguishing hypothetical and absolute necessity: fatalism would require that all truths are absolutely necessary, internally or essentially necessary, but some truths are merely hypothetically necessary (or contingent, as Wolff also puts it), which means necessary as a result of other things. This hardly seems to exonerate Wolff, since if a soldier has deserted not on account of his own nature but due to something external, then we might think that he is even less responsible for the desertion. Leuenberger argues, however, that Wolff's account of the principle of contradiction, developed independently of his explicit modal doctrines, might just be enough to get him off the hook.

Wolff was undoubtedly important for Kant's reception of Leibniz's views, but in the third contribution to the volume, Ohad Nachtomy focuses on Leibniz's and Kant's accounts of both existence and possibility in order to tell some of the complex story of what exactly has changed between them. Leibniz defines possibility as conceivability without contradiction in the divine mind, and thus gives strong expression to the idea that mere possibilities exist without ever becoming actual, an idea which transcends the Aristotelian statistical or temporal account of modality, according to which the actual exists now, the possible at some time, and the necessary at all times. Divine and self-consistent conceivability, as Nachtomy shows, involves a combinatorial and ultimately actualist account of possibility: there are purely actual ultimate simple constituents that when combined in consistent ways constitute possibilities, and Leibniz identifies these atomistic simples with God's attributes. As for existence, God's is distinct from that of his creatures: in his modified version of Anselm's ontological proof, existence is treated as a perfection and predicate, whereas in created things existence is, Leibniz holds, a matter of being compatible or compossible with the most perfect series of things that God, by virtue of his goodness, has actualized. Kant's critique of the ontological argument, according to the claim that existence is not a predicate is, then, Leibnizian in spirit insofar as it merely extends his view that existence is not a predicate in created things to the creator himself. Moreover, Kant's sole proof for the existence of God in the 1763 *Beweisgrund* essay draws on Leibniz's actualist conception of possibility. To put it very briefly, this proof states that the mere possibility of something, of anything at all, requires something actual and necessary, namely God.

Jessica Leech addresses the reflection on possibility in the mature Kant's critical philosophy, and draws our attention to a passage of the 'Ideal of Pure Reason', in the 'Transcendental Dialectic', where Kant seems to criticize his own earlier proof for the existence of God in the *Beweisgrund* essay. In asking how 'reason comes to regard all the possibility of things as derived from a single possibility, namely that of the highest reality', however, Kant also seems to speak of a material and not just formal condition of possibility. Earlier in the first *Critique,* in the chapter of the 'Analytic of Principles' entitled 'The Postulates of Empirical Thinking in General', real possibility—as distinct from logical possibility—was defined as what agrees with the formal conditions of experience. But in the Dialectic, Kant supplements this analysis with the idea that nothing can be given to us unless it is given within a sum total of empirical reality. In this way, Leech argues, Kant develops the idea that real possibility concerns the cognition, which involves sensibility, and not just the thought of an object.

In 'Hegel's Expressivist Modal Realism' Christopher Yeomans illuminates Hegel's distinctions between logical, real and absolute modality. Hegel thinks of real modality—in distinction to the logical possibility governed by the principle of contradiction—in a material sense as the manifold circumstances conditioning a present actuality. Hegel enthusiastically endorses the position that real possibility is but reality or actuality: the possibility of a particular fact, say a car winning the race, is grounded in other actualities, say the quality of the engine and tyres; and the relation between these actualities can be characterized according to an idea of causal necessity. The Hegelian modal notions thus are by no means exclusively characteristics of concepts and judgements, and are characteristic of events themselves. From Hegel's perspective of absolute modality, however, if we add more detail to any given modal description, then this does not simply add more detail to our knowledge of the real possibilities that necessarily cause the present actuality. Rather, as Yeomans puts it, such addition 'only draws the actual and the possible together into the bonds of necessity, and in so doing renders the possible more an internal *feature* of the actual rather than an external *condition on* it'. With this change of philosophical perspective from the real to the absolute, the significance of the three modal terms has changed: 'actuality is the whole pattern of variation or course of development rather than the merely existent states; possibilities are the specific contents of that pattern or phenomena in that course (i.e. what were formerly thought to be actualities) rather than external conditions; and necessity is the structure of that pattern or the force of that development'. Yeomans shows how this realist conception of the modalities as fundamental aspects of the world is a function of what he terms an 'expressivist realism', according to which the absolute actuality posited by Hegel is the self-altering process of its own manifestation.

Thomas Baldwin offers a synoptic account of Russell's changing views concerning possibility and necessity. An intuitionist view of logical necessity, according to which it is a fundamental, indefinable property that is 'purely and simple perceived', swiftly gives way to scepticism concerning whether necessity exists at all. It cannot be explained by analyticity, as Russell argues in 'Necessity and Possibility', which he read to the Oxford Philosophical Society in 1905, since some propositions are felt to be necessary when they are not analytic; '*bad* does not mean the same as *not-good*, and therefore mere logic will never prove that *good* and *bad* are any more incompatible than *round* and *blue*'. Necessity does not have the same extension as analyticity, and Russell is tempted by the idea that it is grounded on the universal truth of the relevant propositional function. Russell seems in this way to return to an Aristotelian statistical conception of modality, according to which the necessary is that which is true at all times. Baldwin shows how Russell's thinking mirrors David Hume's famous sceptical construal of causal necessity: Hume attempts to account for causation in terms of the constant conjunction of events and a psychological feeling of necessity, and, similarly, Russell is led towards the position that the *feeling* of necessity is grounded in the propositional function's constantly and universally being true. In his later work, in the *Problems of Philosophy* and elsewhere, however, Russell often suggests that the domain of quantification of propositional functions is possible worlds—the idiom was familiar to him from his early book on Leibniz. Yet he is clearly reluctant to admit that these possible worlds have any kind of non-actual reality, and thus, as Baldwin shows, his commitments point towards what in contemporary modal theory would be called a quasi-linguistic modal ersatzism.

In 'Modality and Degrees of Truth: An Austro-Polish Sideline in Twentieth-Century Modal Thought', Peter Simons explores what he takes to be a third-way in construing modality that rejects both linguistic accounts and the poly-cosmism of possible world theory. Simons examines two versions of this third-way advanced by Austrian Alexius Meinong—whose theory of objects that do not exist, but which are no less in being, occupied Russell for some time—and the Polish logician Jan Łukasiewicz respectively. Some of Meinong's non-existent objects are *incomplete*, and this allows him in 1915 to account for objective probability (which he names possibility) with an idea of degrees of truth: the proposition 'My draw of a card from the pack tomorrow will be a king' is neither simply wholly true nor wholly false, and this regardless of the draw that I will actually make tomorrow, but has a degree of truth, a degree corresponding to the proportions of kings in a pack of cards, between 0 and 1. This scale ends in truth rather than necessity, but Meinong is nevertheless led to the idea that some statements are true in such a way that their truth is written into them as part of their very nature, and are thus necessary. Łukasiewicz, the inventor of fuzzy logic, visited Meinong in Graz in 1908 and 1909, and in 1913 published his own work on probability, according to which some propositions are *indefinite* and have truth values between and 0 and 1; and in 1917 he began to extend this analysis to *definite* propositions about future contingencies. In conclusion, Simons shows how Łukasiewicz and Meinong's attempts to provide a logical and ontological basis for probability theory have a precedent in the work of Bernard Bolzano.

In my contribution, 'Heidegger on "Possibility"', I examine how an idea of possibility is crucial to Martin Heidegger's philosophical project, and what it means for him to claim, in inverting Aristotle's dicta, that possibility is not 'lower' but rather 'higher' than actuality, which is to say not less but more in being than actuality. The most profound sense of the possible, for Heidegger, is not possibility as constituting a realm of possibilia contrasting with actuality, and it is not conceptual possibility. For the being of what we call 'actually existing' things is already a being-possible. The stuff of daily life—door handles, chairs, and, of course, Heidegger's famous hammer—affords possibilities for action in a given situation, but for all that these possibilities are relative to the agent, they are understood, Heidegger argues, as belonging to things prior to any act of judgement or theoretical cognition. Things *are* a function of how we pragmatically, contextually, and pre-theoretically understand them, and if by the actuality of a thing we mean its being present before the eyes as an isolated object (this is, loosely, what Heidegger takes 'actuality' to mean), then things are not, at least as we ordinarily and primarily encounter them, actual at all. Heidegger's claim that possibility is superior to actuality is, however, developed most fully in his account of the being that we are (*Dasein* in Heidegger's German), an account based, as I show, on an interpretation of Aristotle's definition of movement as the 'actuality of the possible as possible'. Heidegger is not alone among twentieth-century Aristotelians in reading this definition of movement as announcing a mode of being where possibility or potentiality really shows itself and fully exists as the possibility that it is; but he argues that movement as a mode of being essentially characterizing *Dasein*, in its care and striving, cannot be reduced to traditional notions of actuality. In the later sections of the essay, after determining how against this background Heidegger's famous analysis of death as the 'possibility of impossibility' can be understood, I explore how his (neo)-Aristotelian claims are modified in the 1930s by a reflection on art-production. Heidegger's thinking at this time in many ways turns on a revised interpretation of Aristotle's modal metaphysics, and understanding this, I contend, is crucial to any engagement with his later work.

Finally, in 'De Re Modality in the Late Twentieth-Century: The Prescient Quine', John Divers both clarifies and invites greater appreciation of Willard Van Orman Quine's scepticism concerning de re modality. On the basis of his demonstration—with his famous number of planets paradox—of the referential opacity of de re modal predication, Quine, Divers contends, neither argues that de re modal predication is absolutely unintelligible, nor makes an unwarranted assumption that the commitments necessary to defend its intelligibility involve unbearable costs.

The commitments amount to: 1) the idea of a language-independent modality and thus to some kind of metaphysical doctrine of Aristotelian essentialism; 2) a logic of variables and singular terms that is more complicated and weaker than the orthodox approach involved in classical first order logic; and 3) the principle of the necessity of identity as a thesis of quantified modal logic. Quine's presumption that the cost of these commitments could not be borne was not, in context, unwarranted because, Divers contends, his dialectical opponents were the logical empiricists who would have to count them as unbearable. Divers subsequently assesses how the modal philosophies of Saul Kripke, David Lewis, and Kit Fine fare in relation to Quine's predictions. Kripke enthusiastically accepts all three commitments, whereas Lewis and Fine refuse the 'fundamentalist' presumption about de re modal predication on which Quine's predictions are based. Quine, Divers contends, would not have been surprised by their ideas that de re modal predication might be defended as a non-canonical or secondary aspect of 'total theory', and would have been surprised only by the extraordinary metaphysical lengths to which both philosophers are prepared to go in prosecuting this strategy.

Early drafts of a majority of the essays were presented at the 2013 British Society for the History of Philosophy annual conference, but two new essays were commissioned subsequently in order to extend the philosophical and historical range of the volume. A volume such as this cannot pretend to offer a complete survey—but the sense in which any such survey could be complete is open to question—of modal metaphysics in modern philosophy. There are some important modern philosophers who do not directly feature within it, but the volume is intended to act as a spur to further work in the history of modal metaphysics. Given that the last collected volume on modality in modern philosophy, the excellent *Modern Modalities: Studies of the History of Modal Theories from Medieval Nominalism to Logical Positivism*—edited by Simo Knuuttila, and featuring essays on Descartes, Hobbes, Leibniz, Hegel, nineteenth-century French and British philosophy, Frege and Logical Positivism—is nearly thirty years old and relatively hard to come by, it was time for a new collected volume in the field. The present collection complements Knuuttila's volume nicely, and by no means renders it obsolete.

## Bibliography

Blackburn, S., 'Morals and Modals' in *Essays in Quasi-Realism* (Oxford: Oxford University Press, 1993), 52–74.

Divers, J., *Possible Worlds* (London: Routledge, 2002).

Jacobs, J. D., 'A Powers Theory of Modality; or, How I Learned to Stop Worrying and Reject Possible Worlds', *Philosophical Studies* 151: 227–48.

Knuuttila, S. (ed.), *Reforging the Great Chain of Being* (Dordrecht: D. Reidel, 1980).

Knuutila, S. (ed.), *Modern Modalities: Studies of the History of Modal Theories from Medieval Nominalism to Logical Positivism* (Dordrecht: Kluwer, 1988).
Knuuttila, S., 'Modality' in *The Oxford Handbook of Medieval Philosophy*, edited by J. Marenbon (Oxford: Oxford University Press, 2012), 312–41.
Lovejoy, A., *The Great Chain of Being* (Cambridge MA: Harvard University Press, 1936).
Vetter, B., *Potentiality* (Oxford: Oxford University Press, 2015).

# 1

# Aspects of Spinoza's Theory of Essence

## Formal Essence, Non-Existence, and Two Types of Actuality

*Mogens Lærke*

Over the last decade, Spinoza scholarship has witnessed a subtle but forceful shift towards Platonizing interpretations.[1,2] By such interpretations I understand what Anthony Kenny, referring to Descartes, has described as 'Platonism about essences',[3] or interpretations that ascribe to essences a kind of being different from the kind of being that pertains to existences. According to Valtteri Viljanen, for example, 'Spinoza's thought belongs to—and may in fact be the highpoint of—an essentialist tradition that originates with Plato and functions as a shared background for the scholastics....'[4] Emanuela Scribano similarly argues that 'in Spinoza's metaphysics, the

[1] I am grateful to Oberto Marrama, Andrea Sangiacomo, Tad Schmaltz, Steven Nadler, John Brandau, Julie Klein, Raphaële Andrault, Ohad Nachtomy, Sean Winkler, Karel D'huyvetters, and Valtteri Viljanen, who all provided extensive comments, criticisms, and corrections that helped me improve this paper substantially.

[2] For Spinoza's works, I have used the following editions: *Opera*, edited by C. Gebhardt, Heidelberg: Carl Winter Verlag, 1925, vols I–IV, and *The Collected Works of Spinoza*, vol. I., edited by E. Curley (Princeton: Princeton University Press, 1985). I use the following abbreviations: E = *Ethics* [roman numerals I–V = Part; D = Definition (when preceding a P); A = Axiom; Exp. = Explication; P = Proposition; D = Demonstration (when following a P); C = Corollary; S = Scholium; e.g. EIP17D is the demonstration of proposition 17 in the first part of the *Ethics*]. For ease of reference and sake of uniformity, when different, I change the nomenclature in quotations from other secondary literature to fit this format. TIE = *Tractatus de intellectus emendatione* [I refer to the § numbers established by Bruder, also employed by Curley]; CM = *Cogitata metaphysica* [following by part and chapter number]; KV = *Korte Verhandeling van God, de Mensch en Deszelfs Welstand* [followed by part, chapter, and section number; the § numbers used by Gerhardt and Curley are used for the appendixes]. Descartes: AT = *Oeuvres de Descartes*, edited by C. Adam and P. Tannery (Paris: Vrin/CNRS, 1964–1976); CSM = *The Philosophical Writings of Descartes*, trans. J. Cottingham et al. (Cambridge: Cambridge University Press, 1984).

[3] A. Kenny, 'The Cartesian Circle and the Eternal Truths', *The Journal of Philosophy* 67:19 (1970): 685–700, p. 696.

[4] V. Viljanen, *Spinoza's Geometry of Power* (Cambridge: Cambridge University Press, 2011), p. 15.

double aspect of the existence of finite modes reproduces the traditional Platonic theory according to which time is an imitation in the sensible world of a more true reality that is outside time.'[5] A Platonizing interpretation of Spinoza is thus an interpretation that allows for an equivocation in Spinoza on the term 'existence', or, as Charles Jarrett puts it, following a suggestion already made by Albert Rivaud in his 1905 monograph *Les notions d'essence et d'existence dans la philosophie de Spinoza*, an interpretation according to which 'Spinoza employs what might be called a "dual notion" of existence.'[6]

Recent discussions have largely turned on Spinoza's use of the notions *essentia formalis* and *essentia actualis*. For some Platonizing commentators, formal essences and actual essences are separate essences. For others, actual essences are actualized formal essences. In both cases, however, actual essences are nothing but actually existing essences, and formal essences as such do not imply actual existence.[7] Formal essence is seen either as a being distinct from actual essence, or as a different kind of being of essence than the actual one, in both cases giving rise to a multi-layer ontology where, for each thing, there is, on the one hand, a being of its essence (or formal essence) and, on the other, a being of its existence (or actual essence.) One way of fleshing out this distinction is then to maintain that formal essences are eternal in accordance with Spinoza's notion of eternity as 'existence itself, insofar as it is conceived to follow necessarily from the definition alone of an eternal thing' and which 'cannot be explained by duration or time' (EID8). Actual essences, for their part, have duration, or 'determinate existence' as Spinoza sometimes terms it (e.g. EP21D), i.e. they enter and leave actual existence. In this context, non-actualized formal essences are often associated with the notorious non-existent modes that Spinoza discusses in EIIP8. On the Platonizing picture, the ideas that we have of non-existent things, ideas of formal essences, are ideas about a different kind of existence than the ideas we have of actualized essences, i.e. actually existent things. The first ideas are then about eternal formal essences; the second ones are about the actual essences of presently existing things. Some commentators have compared formal essences and non-existing modes with *possibilia*, thus making of the realm of formal essences in important respects comparable to Leibniz's *regio possibilitatis*.[8] Others

[5] E. Scribano, *Guida alla lettura dell'Etica di Spinoza* (Roma-Bari: Editori Laterza, 2008), p. 45. I am grateful to Oberto Marrama who translated the relevant passages from Italian for me.

[6] C. Jarrett, 'Spinoza's Distinction between Essence and Existence', in *Iyyun: The Jerusalem Philosophical Quarterly* 50 (2001): 245–52, p. 247. For Rivaud, see *Les notions d'essence et d'existence dans la philosophie de Spinoza* (Paris: Félix Alcan, 1905). For an even older study going down the same road, see L. Busse, 'Über die Bedeutung der Begriffe "'essentia" und "'existentia" bei Spinoza', *Vierteljahrschrift für wissenschaftliche Philosophie* 10 (1886): 283–306, in particular p. 291.

[7] See for example D. Garrett, 'Spinoza on the Essence of the Human Body and the Part of the Mind that is Eternal,' in *Cambridge Companion to Spinoza's Ethics*, edited by O. Koistinen (Cambridge: Cambridge University Press, 2009), 286–7; T. M. Ward, 'Spinoza on the Essences of Modes', *British Journal for the History of Philosophy* 19:1 (2011): 19–46, p. 35; V. Viljanen, 'Spinoza on Virtue and Eternity,' in *Essays on Spinoza's Ethical Theory*, edited by M. J. Kisner and A. Youpa (New York: Oxford University Press, 2014), 264–5.

[8] See A. Donagan, *Spinoza* (Chicago: Chicago University Press, 1988), pp. 58–9; R. J. Delahunty, *Spinoza* (London: Routledge & Kegan Paul, 1985), pp. 295–6; J. Bennett, *A Study of Spinoza's Ethics*

have tried to situate the formal essences within Spinoza's framework of modes, inscribing them in different ways within Spinoza's difficult doctrine of infinite modes.[9] Yet others have given the Platonizing tendency an additional turn of the screw, by maintaining that formal essences are not individual, but general.[10] But all these different readings agree on one point, namely that Spinoza operates with several kinds of existence and that a dichotomy opposing formal to actual essences captures the difference between them.

The various Platonizing interpretations do not form a unified front, but emerge from very different approaches and in various degrees of explicitness. The convergence between them could seem to lend additional plausibility to the basic point they have in common. In the following, however, I want to resist that conclusion. The main reason for my resistance is my conviction that Spinoza, as Gilles Deleuze already pointed out half a century ago, is committed to the univocity of being.[11] In short, for Spinoza, things either exist or they don't, there is no class of beings in between, and existence is always said in the same sense of the things of which it is said. I think there are several ways in which one can arrive at this conclusion. Here is, briefly, one of them. The central tenet of Spinoza's metaphysics is the idea that there is a unique substance, God or Nature, of which all other things are modes (EIP14C1–2; EIP25C; KV II, v, § 10, etc.). All things are necessarily *in* God, as he also puts it (EIP15). Moreover, substance does not exist over and above the production of modes that necessarily follows from it (EIP16), insofar as its essence just is such productive, causally effective power (EIP34). As Spinoza maintains in EIP25S, 'God must be called the cause of all things in the same sense in which he must be called the cause of himself.'[12] This 'same sense clause' posits the existence of modes as grounded in exactly the same causal process as that in which substance is itself grounded: Substance produces the existence of modes as it produces the existence of itself. The existence attributed to modes is *the exact same existence* as the one

(Indianapolis: Hackett, 1984), pp. 357–8; W. Matson, 'Body, Essence and Mind Eternity in Spinoza', in *Spinoza: Issues and Directions*, edited by E. Curley and P.-F. Moreau (Leiden: E. J. Brill, 1990), pp. 88–9. E. Yakira suggests that we should distinguish the level of 'existence' from that of 'ontology' and claims that nonexistent modes are 'in some sense... just possibilities' ('Ideas of Nonexistent Modes: Ethics II Proposition 8, its Corollary and Scholium', in *Spinoza on Knowledge and the Human Mind*, edited by Y. Yovel and G. Segal, (Brill: Leiden, 1994), pp. 160 and 164).

9 There is, however, little consensus about *how* formal essences relate to infinite modes. For Martin, Garrett, and Ward, all formal essences are infinite modes. For Scribano and Mignini, the total series of formal essences constitutes a *mediate* infinite mode. According to Steven Nadler, in an insightful and non-Platonizing reading, the total series of formal essences constitutes an *immediate* infinite mode. For Nadler, see 'Spinoza's Monism and the Reality of the Finite,' in *Spinoza on Monism*, edited by P. Goff (New York: Palgrave Macmillan, 2012), p. 234. For the others, see later in this chapter for references.

10 See later in this chapter, on Ward, Garret, and Martin.

11 See G. Deleuze, *Spinoza et le problème de l'expression* (Paris: Minuit, 1968), pp. 44–58.

12 This is, of course, the doctrine commonly known as Spinoza's 'monism'. I have however argued elsewhere that this is not an appropriate description of Spinoza's position, on both historical and doctrinal grounds. See M. Lærke, 'Spinoza's Monism? What Monism?', in *Spinoza on Monism*, edited by P. Goff (New York: Palgrave Macmillan, 2012), 244–61.

attributed to substance, but modified.[13] Consequently, existence is always said in the same sense about the things of which it is said, be it substance or modes. As Spinoza puts it in EIIP45S, we here touch upon 'the very nature of existence, which is attributed to singular things because infinitely many things follow from the necessity of God's nature in infinitely many modes.'

Within Spinoza's ontological framework, therefore, it is hard to see how there could be room for distinguishing between the being of essences and the being of existences as two different ontological 'levels' inhabited by ontologically distinct beings, or by beings in ontologically distinct states, such as Platonizing interpretations suggest, since there is in Spinoza only one possible ontological level, namely that of modified substance, 'being in God', or 'existence itself' as Spinoza calls it in EID8. The theory of the divine attributes will do no work in alleviating this problem, since Spinoza's commitment to inter-attribute parallelism, or ontological parallelism, requires that all attributes include the same modes, or that all attributes are inhabited by the same things in the same order.[14]

This is the most general reason for my reticence towards Platonizing readings. In itself, it is obviously insufficient to discard them. After all, there could very well be a real tension in Spinoza's metaphysics on this point. Or, as another option, Spinoza's theory of formal essences might just be proof that my initial intuition about the univocity of being in Spinoza is in fact mistaken. In the following, I will however identify the root of Platonizing readings and establish exactly where they go wrong and why. I principally argue that the central opposition between formal essences and actual essences is a false dichotomy. Spinoza never intended to juxtapose those terms and the use Platonizing readings have made of them relies on a misconstrual which is inconsistent with Spinoza's use of the terminology. In reality, when correctly construed, Spinoza's conception of essence and existence, of formal essences, existent and non-existent modes, actual essence and actual existence, all converge towards a single, strongly anti-Platonist conception according to which everything that is and that can be conceived is and is conceived to be on one and the same ontological level, namely that of modified substance.

In the following, I thus provide, against Platonizing readings, the basic outline of an alternative 'aspectual' reading of Spinoza's doctrine of formal essence, objective being, existence and non-existence, and actuality, that conforms to his one-level ontology. By an aspectual reading, I understand a reading that takes all these different qualifications to always express different aspects of one and the same thing rather than different entities.[15] I proceed in the following manner. First, I present a series of

[13] On this point, see my 'Spinoza and the Cosmological Argument According to Letter 12', *British Journal for the History of Philosophy* 21:1 (2013): 57–77.

[14] On the varieties of parallelism in Spinoza, see note 42, this chapter.

[15] I am not the only one to press such an 'aspectival' reading of Spinoza's metaphysics. Julie Klein, in an article on the eternity of the mind of particular relevance to us, strongly emphasizes how 'both Spinoza and Gersonides rely on an account of perspectival or conceptual, as opposed to real, difference', pointing to

Platonizing readings, indicating for each of them where I think they conflict with central tenets of Spinoza's philosophy. In the following section, I turn to my own analysis of the various terms in play, taking my point of departure in basic, lexical analyses of Spinoza's use of the relevant terms in the *Ethics*, placing each term in its proper context within Spinoza's deductive framework, while determining the often polemical motivations that lead him to introduce them. In the following section on formal essences, my main objective is to show that, in Spinoza's conceptual universe, formal essence is not opposed to actual essence, but rather to objective being. Next, I turn to Spinoza's conceptions of existence and non-existence, and how they relate to formal essence. It is my aim here to establish that Spinoza never intended non-existent modes to be construed as un-actualized formal essences. Admittedly, Spinoza clearly asserts that the formal essences of non-existing things are 'contained' in the divine attributes, in order to assure that our adequate ideas of things that currently do not exist can be true. However, to avoid positing two kinds of existence—an actual existence and a non-actual existence—, I argue that non-existing things are contained in the attributes *qua* non-existent through those other existent things that currently are the precise cause or reason of their non-existence. In section 1.5, I establish how Spinoza's notion of 'actual essence', which he identifies with a finite mode's power, or *conatus*, does not imply that it is possible to conceive of an essence that is not actual. Spinoza simply uses the qualification 'actual' to stress the fact that essences are never something merely potential, but always powers *in actu*. When Spinoza speaks of formal essence and actual essence, he is not referring to two entities or even to the same entity in distinct ontological states, but to two eternal aspects of one and the same essence, namely form and power. Consequently, things that do not exist cannot be construed as un-actualized formal essences, and it is misleading to speak of the transition into durational existence of things as an 'actualization' of their essence. In the penultimate section, before concluding, I discuss the notion of actuality in more detail. I show how Spinoza uses the notion of actuality in two different ways when qualifying essences and existences, thus giving rise to an equivocation on the term 'actuality'

Spinoza's constant use of the adverb *quatenus* 'to mark such shifts in perspective.' Such shifts, indicating 'differences of respect', permits one to 'avoid the proliferation of real entities' and express how 'an actual thing can be experienced in irreducibly different ways without violating its ontological integrity; respectival difference holds together what real difference would divide' (see J. R. Klein, '"Something of it Remains": Spinoza and Gersonides on Intellectual Eternity', in *Spinoza and Medieval Jewish Philosophy*, edited by S. Nadler (Cambridge: Cambridge University Press, 2014), 307–29, p. 319). She specifically adopts a notion of 'aspectival difference' to designate the two ways of conceiving things as 'actual' that Spinoza envisages in EVP29S (for similar, 'aspectival' formulations, see Nadler, 'Spinoza's Monism and the Reality of the Finite', in *Spinoza on Monism*, edited by P. Goff (New York: Palgrave Macmillan, 2012), p.228). Yitzhak Melamed also appeals to a notion of 'aspects' when analysing how attributes relate to substance and when analysing the 'multifaceted' nature of ideas in Spinoza (see *Spinoza's Metaphysics: Substance and Thought* (Oxford: Oxford University Press, 2013), pp. xix–xxi and 148–56, and 'The Building Blocks of Spinoza's Metaphysics: Substance, Attributes, and Modes', in *Oxford Handbook to Spinoza*, ed. M. Della Rocca (Oxford: Oxford University Press [forthcoming]).

that becomes explicit when Spinoza himself distinguishes between two types of actuality in EVP29S.

## 1.1 Platonizing Spinoza

Let me first give some examples of Platonizing readings, indicating in each case where I believe they go wrong.

In a recent paper entitled 'The Framework of Essences in Spinoza's *Ethics*', Christopher Martin argues that Spinoza adopts a double framework of essences, allowing not only for singular essences, but also for more general essences, so that things somehow are endowed with two essences, a singular and a general one. Martin maintains that the actual essences and formal essence are separate essences. Formal essences are mediate infinite modes and mediate infinite modes are like laws of nature. Actual essences are 'instances or exemplifications of these laws'.[16] The formal essences are eternal and 'identically exemplified in a number of modes'.[17] The actual essences are durational and particular. The relation between formal and actual essences is thus 'analogous to Plato's characterization of the relation between form-copies and their forms.'[18] This is the most extreme example of a Platonizing reading I have come upon. However, a crucial part of Martin's textual evidence in favour of the idea that Spinoza endorses a notion of 'species-essences', i.e. formal essences common to several individuals, is drawn from a passage in EIP17S, according to which individuals that agree with regard to their essence are such that 'if the essence of one could be destroyed, and become false, the other's essence would also be destroyed, and become false.'[19] The passage is taken as evidence that things have essences in common, so that there are general 'species-essences' not unlike Platonic forms. Martin fails to realize a crucial point. As Alexandre Koyré clearly shows in a

[16] C. P. Martin, 'The Framework of Essences in Spinoza's *Ethics*', *British Journal for the History of Philosophy* 16:3 (2008): 489–509, p. 507.

[17] *Ibid.*, 496. Don Garrett makes a similar suggestion in 'Spinoza on the Essence of the Human Body', p. 291n.

[18] Martin, 'The Framework of Essences', p. 507.

[19] Martin, 'The Framework of Essences', p. 495. In order to defend his claim regarding 'species-essences', Martin moreover draws on EIP8S2 according to which a true definition of man does not involve any specific number of actually existing men in order to conclude that 'the definition of human nature itself is the same for each member of the kind, as it expresses their nature or essence, so is this the same in each instance' (p. 494). The argument is a little quick in concluding that an adequate definition of a thing necessarily expresses the essence of all the individuals falling under that definition. Definitions based on common notions, also described by Spinoza as 'Reason' or the 'second kind of knowledge', do not capture essences at all (see EIIP37 and EIIP44C2). They capture what is common to the multiple essences of the modes that fall under that notion. Hence, the 'common notion' of man is a notion of what is common to the essences of all the individuals that fall under that notion. But that certainly does not involve that that notion is a notion of any single essence. Intuitive, third-kind knowledge does indeed involve ideas and definitions of singular essences, but nothing indicates that such knowledge is about anything but individuals, or that *several* individuals could fall under the essences intuited in third-kind knowledge. Quite to the contrary, intuitive knowledge is knowledge of the essences of *singular* things (see EVP36S).

classic 1950 article, followed by Martial Gueroult, Ferdinand Alquié, and a host of other commentators, Spinoza's development in EIP17S is essentially a *reductio ad absurdum* against the view that 'will and intellect pertain to the eternal essence of God.'[20] The passage Martin brings to the table as evidence is an important part of the, for Spinoza, unacceptable consequences to which this mistaken conception of God's essence will lead. The view regarding essences that he highlights in order to bolster his reading is *not* a view that Spinoza embraces, but exactly one of the consequences that Spinoza deems sufficiently *absurd* to justify rejection of the premises it relies on!

I turn to my second example. In a paper entitled 'Spinoza on the Essence of the Human Body and the Part of the Mind that is Eternal' Don Garrett aims at clarifying Spinoza's conception of the eternity of the mind. Garrett bases his analysis on the following central development:

> ...we must clarify the ontological status of the formal essences and of the ideas that are *of* them. Spinoza strongly implies that formal essences are truly something in their own right... formal essences of singular things must be modes of God.... Every mode... of God is either (i) infinite and eternal, following from God's 'absolute nature'...or (ii) finite and determinate (i.e.) limited in its existence.... But if the formal essences of singular things are modes of God, they can hardly be finite modes. Because they have their own being of existence contained in the attributes of God regardless of when or whether the corresponding singular things themselves exist, it is hard to see why or how they could ever come into or go out of existence, as finite modes do.... The formal essence of a singular thing is thus not identical to the singular thing itself—for the singular thing, having 'a finite and determinate existence' (by E2D7), is a finite mode, whereas the formal essence is an infinite mode... As we have also noted, however, the formal essences of things do somehow ground the actualizability of singular things themselves. From these various clues, we can infer what the formal essence of a singular thing must be: it is the omnipresent modification of an aspect of an attribute of God that consists in the attribute's general capacity to accommodate... the actual existence of a singular thing of the given specific structure whenever and wherever the series of actual finite causes should actually determine it to occur.[21]

The central claim in Garrett's reading is the suggestion that formal essences are infinite modes.[22] Spinoza of course says no such thing but, Garrett contends, 'their [i.e. formal essences'] status as infinite modes is strongly confirmed in EVP23S by

[20] See A. Koyré, 'Le Chien, constellation céleste, et le chien, animal aboyant', *Revue de métaphysique et de morale* 55:1 (1950): 50–9; M. Gueroult, *Spinoza I: Dieu* (Paris: Aubier Montaigne, 1968), 272–95; F. Alquié, *Le Rationalisme de Spinoza* (Paris: Presses universitaires de France, 1981), 152–6. On the reading and misreading of EIP17S, see also M. Lærke, *Leibniz lecteur de Spinoza. La genèse d'une opposition complexe* (Paris: Honoré Champion, 2008), 763–79, and 'Leibniz on Spinoza's *Tractatus de intellectus emendatione*', in *The Young Spinoza. A Metaphysician in the Making*, edited by Y. Melamed (Oxford: Oxford University Press, 2015), 106–20. An English translation by O. Marrama of Koyré's classic article, including a helpful introduction, has been published as "The Dog that is a Heavenly Constellation and the Dog that is a Barking Animal by Alexandre Koyré", in *The Leibniz Review* 24 (2014): 95–108.

[21] Garrett, 'Spinoza on the Essence of the Human Body', pp. 289–90.

[22] Martin and Ward also espouse the view that formal essences are infinite modes. See Martin, 'The Framework of Essences', p. 496n, and pp. 500–4; and Ward, 'Spinoza on the Essences of Modes', p. 31.

Spinoza's description of the parallel "idea, which expresses [i.e. is of] the essence of the body" as "a mode of thinking... which is necessarily eternal."'[23] This subsequently leads him to conclude that the part of the mind that is eternal coincides with its formal essence, and that what is eternal in the mind is an infinite mode. This complicated and somewhat obscure analysis relies heavily on the idea that the formal essences, as Garrett writes, 'have their own being of existence' that is 'not identical to the thing itself'. This point is argued on the basis of EVP21–22, i.e. the two propositions where Spinoza considers the difference between recollections and memory and what is eternal in the mind. Garrett writes about those two propositions that, 'it is clear from the demonstrations of these two consecutive propositions that Spinoza is invoking a distinction of some kind between the actual essence of a human body and the formal essence of a human body.'[24] On verification, however, EVP21&D mentions only the 'actual existence' of the body[25] and EVP22&D only the 'essence' of the human body.[26] But Spinoza at no point mentions 'actual essences' or 'formal essences' in these two propositions.[27] Garrett's analysis thus rests on very little textual evidence.[28]

As a third example, Thomas M. Ward suggests that we should understand Spinoza's treatment of formal essences in the *Ethics* in light of the use of that same notion in the *Cogitata metaphysica*, a text published in 1663 as an appendix to Spinoza's geometrical exposition of Descartes's philosophy. According to Ward, Spinoza posits 'an abstract realm of essences'.[29] Indeed, 'essences, or what [Spinoza]

[23] Garrett, 'Spinoza on the Essence of the Human Body', p. 290.

[24] Garrett, 'Spinoza on the Essence of the Human Body', p. 285.

[25] See EVP21D: 'The Mind neither expresses the actual existence of the Body, nor conceives the Body's affections as actual, except while the Body endures...'.

[26] See EVP22: '[I]n God there is necessarily an idea that expresses the essence of this of that human body...'.

[27] It could be objected that EVP21D contains an indirect reference to formal essence, on account of the reference to EIIP8C that it contains. EVP21D, however, only uses EIIP8C in order to establish that the mind does not conceive 'the Body's affections as actual, except while the body endures (by EIIP8C).' But this claim is exclusively about *actual existence* in duration. It does not in itself involve any claim about formal essences, let alone non-actual formal essences.

[28] Moreover, there is something very puzzling about the conclusion that formal essences are infinite modes. When Spinoza speaks of infinite modes, mysterious as they admittedly are, they are identified as something entirely different, namely as 'movement and rest' and the 'infinite intellect' when it comes to the immediate infinite modes, and 'the face of the whole universe' when it comes to the mediate infinite modes (see Letter 64 from Spinoza to Tschirnhaus; for helpful, and very different, surveys of Spinoza's theory of infinite modes, see T. Schmaltz, 'Spinoza's Mediate Infinite Mode', in *Journal of the History of Philosophy* 35:2 (1997): 199–235; Y. Melamed, *Spinoza's Metaphysics: Substance and Thought* (Oxford: New York, 2013), 113–36). But I see no discernible reason why such modes would qualify as formal essences or *vice versa*. Apparently, we will have to completely revise our conception of what infinite modes *are*, and of what Spinoza *says* they are, in order to make that notion a viable *explanans* for the notion of formal essence. In other words, the *explanandum* is only explained by the *explanans* on the condition that we put to one side the only actually available (i.e. explicitly stated, by Spinoza) account of what the *explanans* is. From there on, we are just groping around in the dark, because we attempt to elucidate a notion A that we seek to understand by means of another notion B of which we no longer know the meaning.

[29] Ward, 'Spinoza on the Essences of Modes', p. 32.

sometimes calls formal essences, are produced by the divine essence prior to and independent of the creation of finite modes, and according to which essences are formal or exemplar causes of finite modes.'[30] Ward thus supports the notion that formal essences exist separately in God prior to existences as their 'exemplars' in a way reminiscent of Martin's 'species-essences.'[31] Ward mainly draws on the *Cogitata metaphysica* in order to reach that conclusion. In a central passage Spinoza writes regarding formal essences:

> [T]he formal essence neither is by itself nor has been created, for both those presuppose that the thing actually exists. Rather it depends on the divine essence alone, in which all things are contained. So in this sense we agree with those who say that the essences of things are eternal. (CM I, ii)

Now, generally speaking, the *Cogitata metaphysica* cannot be taken to express Spinoza's own position. Spinoza's stated purpose is not to present his own philosophy, but 'to show that the common Logic and Philosophy only serve to train and strengthen the memory...; but these disciplines do not serve to train the intellect.' He cautions that his 'intention here is only to explain the more obscure things which are commonly treated by Writers on Metaphysics' (CM I, i). It is a text intended for Cartesians who wish to form an idea of the kind of scholasticism taught in seventeenth-century Dutch universities. There are, however, passages in the CM where Spinoza does indeed refer to his own views, and the passage from CM I, ii, quoted above, cannot be that easily discarded, on account of the phrase 'we agree with those who say'. Spinoza clearly tries to express his own views within the 'metaphysical' framework he is in the process of elucidating. However, what Spinoza subscribes to in this specific passage is, first, the idea that the formal essence of things depends on the divine essence in which all things are contained. This is nothing over and above what he explicitly says in EIIP8 as well. What he rejects, on the contrary, is the idea that formal essences are uncreated and exist *per se*, independently from God. But this, again, does not shed any additional light on the *Ethics*, where he also clearly affirms in EIP25 that 'God is the efficient cause, not only of the existence of things, but also of their essence', on the grounds that 'nothing can be conceived without God.'[32] Most importantly, nothing in these passages implies the stronger claim that formal essences are 'prior to and independent of' actual existences. On the contrary, Spinoza's polemical target in EIP25 seems *prima facie* to be a particularly strong type of Platonism where essences have being independent even from God. In order for Ward's interpretation to have any plausibility, we need textual evidence in favour of the priority and independence of formal essences. *This* view, however, is developed

[30] Ward, 'Spinoza on the Essences of Modes', p. 19. See also p. 20: 'Spinoza... thinks of essences, or what he sometimes calls *formal essences*, as produced by the divine essence prior to and as the formal of exemplar causes of finite modes', and p. 36: 'Spinoza seems to posit essences as existing in some way prior to and independent of the existence (with duration) of particular things, and as produced by God.'

[31] Ward, 'Spinoza on the Essences of Modes', pp. 33–4.

[32] On this point, see also EIIP10S.

nowhere in the *Ethics* and only developed in the *Cogitata metaphysica* in passages where Spinoza presents doctrines he is in fact fiercely opposed to. Hence, in the beginning of CM II, i, Spinoza refers to a doctrine according to which 'God contains eminently what is found formally in created things, i.e., . . . God has attributes in which all created things are contained in a more eminent way' (CM II, i). It is in the context of such a traditional doctrine of creation by a transcendent (or 'eminent') God that one also must conclude that the '*being of Essence* is nothing but that manner in which created things are comprehended in the attributes of God. *Being of idea* is spoken of insofar as al things are contained objectively in God's idea. . . . Finally, *being of Existence* is the essence itself of things outside God, considered in itself. It is attributed to things after they have been created by God' (CM II, i). For Spinoza himself, of course, God is not 'eminent', he does not 'create' things, and there certainly are no 'things outside God'!

Like Ward, Tad Schmaltz turns to the *Cogitata metaphysica* for help. He does however proceed with considerably more caution and sensitivity to the peculiar status of the text. I will quote his analysis at length, since it is both dense and clear as it stands, and does not require much commentary:

It might seem natural, then, to turn to CM for an understanding of EIIP8. To be sure, the relevant discussions in these two texts differ in fundamental respects. For instance, the claim in CM that the essences of non-existent modes are comprehended in the created substances they modify is one the Spinoza of the *Ethics* cannot accept. For the mature Spinoza, there can be no difference between the containment of essences of non-existent modes in the substance they modify, on the one hand, and their containment in the divine essence, on the other. What is relevant to EIIP8, though, is the claim in CM that the formal essence of a thing participates in the atemporal eternity of God's essence and does not presuppose the actual existence of that thing. Likewise, the indication in the corollary to EIIP8 seems to be that the formal essences of non-existent modes contained in the divine attributes have an existence that differs from the existence that the modes have 'insofar as they are said to have duration.' In this respect, the formal essence of a singular thing differs from what Spinoza calls in the *Ethics* the 'actual essence' of that thing. The latter is identified in EIIIP7 with the *conatus* 'by which each thing strives [*conatur*] to persevere in being.' This actual essence cannot exist without the singular thing that perseveres in being, nor can that thing exist without its actual essence. In fact, the singular thing is to be identified with its actual essence, and thus the latter is itself a finite mode. Not so, however, in the case of the formal essence of the singular thing. That essence is present atemporally in an attribute whether or not the singular thing has durational existence. This essence therefore cannot be identified with the enduring singular thing. The suggestion in the *Ethics* is rather that it must be identified with the attribute that contains it. This distinction between formal and actual essences is anticipated in the division in CM between the formal essences of things contained in God's essence, on the one hand, and the essences comprehended in those things as they actually exist, on the other.[33]

[33] T. Schmaltz, 'Spinoza on Eternity and Duration: The 1663 Connection', in *The Young Spinoza. A Metaphysician in the Making*, edited by Y. Melamed (New York: Oxford University Press, 2015), 205–20.

Schmaltz's analysis is built up around Spinoza's conceptions of duration and eternity, arguing that the distinction between the being of essence and the being of existence in the CM is reproduced in the *Ethics*' distinction between eternity (pertaining to formal essence) and duration (pertaining to actual essence, or rather actual existence of essence). He does however not appeal to the notions of eminence and creation that eventually will structure the relation between the being of essences and the being of existences/actual essences in the CM, arguing that, in the *Ethics*, the being of essences must rather 'be identified with the attribute that contains it', though without providing much further help in understanding how such identification is to be understood exactly.

Schmaltz's analysis in many respects resembles the one developed by Viljanen in his insightful *Spinoza's Geometry of Power*. Viljanen holds that '[formal] essences are endowed with their own kind of being and a specific resulting ontological structure',[34] that 'the being of essences is the prime layer of reality itself'[35] and consequently, that there are 'two layers of reality, the eternal and the temporal'.[36] It is also strongly reminiscent of the position of Emanuela Scribano, according to whom EIIP8C posits 'a double existence of things, inside and outside time'.[37] Scribano, however, takes one additional step, assimilating the totality of formal essences to the mediate infinite mode. Hence, according to Scribano, formal essences are contained in the attributes of God as the mediate infinite mode, or *totius facies universi*, which she conceives as 'extension *sub specie aeternitatis*', or 'a great receptacle where all the possible relations of motion and rest are contained *ab aeterno*.'[38] The face of the whole universe is the structure and totality of all formal essences in God that are progressively actualized in actual existence: 'The *facies totius universi*, which follows eternally from God, reduplicates itself into a series of events which occur in duration. Now we know that the singular things whose existence is measured in time are the combination of an eternal essence and an existence in duration....'[39]

The converging and in some respects mutually supportive readings by Schmaltz, Viljanen, and Scribano represent the greatest challenge. They are also the most moderate among the Platonizing readings. Contrary to Martin and Ward, none of them claims that Spinoza endorses general essences, 'species-essences' or 'exemplar essences'. Essences remain singular, such as Spinoza's own basic definition of essences clearly suggests that they are.[40] None of them make the unintelligible claim

[34] Viljanen, *Spinoza's Geometry of Power*, p. 24.
[35] Viljanen, *Spinoza's Geometry of Power*, p. 11.
[36] Viljanen, *Spinoza's Geometry of Power*, p. 21.
[37] Scribano, *Guida alla lettura dell'Etica di Spinoza*, p. 88.
[38] Scribano, *Guida alla lettura dell'Etica di Spinoza*, p. 55.
[39] Scribano, *Guida alla lettura dell'Etica di Spinoza*, p. 55. For a similar analysis, see Spinoza, *Opere*, edited by F. Mignini and O. Proietti (Milan: Arnoldo Mondadori Editore, 2007), pp. 1640–1, n22.
[40] See EIID2: 'I say that to the essence of the essence of any thing belongs ... that without which the thing neither can be nor be conceived, and which can neither be nor be conceived without the thing.' It is of course the last part of the definition that suggests that essences are proper to the things of which they are essences, and therefore singular. See also EIIP10S.

shared by Martin and Garrett that each and every formal essence is an infinite mode. So what is wrong? It here really comes back to the finer details of their interpretations. On a Platonizing reading, however moderate, we must *either* see the formal and actual essence as two really distinct essences *or* see actual essence as the formal essence actualized. Schmaltz, Viljanen, and Scribano are not particularly clear about which option they each favour,[41] so I shall consider both.

In the first case, where formal essence and actual essence are seen as two separate essences, there are yet two other options: *either* positing a transcendent realm of formal essences separate from that of actual essences *or* positing the existence of two separate essences of the same thing in each attribute. The first option is the position Spinoza attributes to the 'metaphysicians' in the CM: essences have a being different from existences because the former are contained eminently in God, while the latter are created and outside God. Such distinctions are obviously not available to Spinoza in the *Ethics*. Both existences and essences must be contained within the same divine attributes and Spinoza clearly affirms that they are. The second option doubles the essential presence of a mode within the divine attribute of which it is a mode. Hence, the pine tree outside my window, say, has a formal essence contained in the attribute of extension that I can grasp adequately as an eternal truth. However, while that tree exists, it also has an actual essence, equally contained in the attribute of extension, but somehow really distinct from the formal essence. Thus, when a thing enters existence, this involves the coming into being of a distinct actual essence which is, as it were, superadded to the formal one, all within the same attribute. Thus, when a thing exists, this thing has two distinct essences, a formal and an actual one, both contained within the same attribute, and combined with each other they form the composite essence of the thing *qua* existent. On the face of it, this is a rather odd theory. More importantly, however, it contradicts Spinoza's ideas-things parallelism, or epistemological parallelism,[42] according to which 'the order of connection of ideas is the same

[41] Schmaltz does not address the problem and the central passage quoted above does not really allow for any conjectures as to his position on the matter. For Scribano, existing things are a 'combination' of an eternal essence and an existence in duration (Scribano, *Guida alla lettura dell'Etica di Spinoza*). This suggests that she embraces the first option. For Viljanen, 'the eternal and infinite systems of essences, together with the determinations specified by them, are thus converted into and correspond to the spatio-durational existence of the actual world' (*Spinoza's Geometry of Power*, p. 32). 'Converted into' here suggests that the second option is right, the 'correspond to' rather that the first one is. He does, however, also refer to a passage in Garrett according to which 'we may also think of the actual essence of a singular thing as the *actualization* or instantiation of its existing formal essence, rendering the thing itself actual' (Garrett, 'Spinoza on the Essence of the Human Body', pp. 286–7). This reference pulls Viljanen in the direction of the second option.

[42] Like the notion of 'ontological parallelism' already employed above, I borrow the notion of 'epistemological parallelism' from Gilles Deleuze (see *Spinoza et le problème de l'expression*, pp. 99–113). While there is ontological parallelism between corresponding expressions of a same mode of substance under different attributes, e.g. thought and extension, there is epistemological parallelism between a mode of substance, or thing, expressed under some attribute, e.g. thought or extension, and the corresponding idea or representation of that mode given in the divine intellect. For other, similar, readings of Spinoza's multiple 'parallelisms', see M. Gueroult, *Spinoza II: l'âme* (Paris: Aubier-Montaigne, 1974), pp. 47–102,

as the order and connection of things' (EIIP7) and 'whatever follows formally from God's infinite nature follows objectively in God from his idea in the same order and with the same connection' (EIIP7C). On this first possible scenario, there is not, as EIIP7 requires, a one-to-one relationship between ideas and things, for the adequate idea of some existing thing is in fact the idea of *two* distinct things, or entities, namely of the formal essence *and* of the actual essence. Consequently, one cannot consider the formal essence and actual essence of a same thing adequately as two separate entities, on pain of violating epistemological parallelism.

According to the second possible scenario, we think about the actual essence as the actualization *of* the formal essence. In this case, there are not two essences combined in the existing thing, but only the one (eternal) formal essence insofar as it is actualized (for some given duration of time). The temptation to reactivate the vocabulary of pure *possibilia* in the context of this reading is very strong. A number of commentators have, in the past, gone in that direction.[43] Scribano also seems to slide into that position when holding that we should think about the formal essences in terms of 'all the possible relations of motion and rest' as they are contained *ab aeterno* in the mediate infinite mode.[44] But conceiving the reality of formal essences in terms of possibility or potentiality, indeed conceiving anything in those terms, is forcefully rejected by Spinoza himself in EIP33S. This temptation must be resisted if there is to be any coherence to Spinoza's overall doctrine. Even a Platonizing reading, however, need not succumb to it. Viljanen, for example, avoids it by confining his reading of formal and actual essence to a formulation in terms of the eternal and the durational, but without ever suggesting that these dichotomies map on to the modal distinction between the possible and the real. This last approach represents my most formidable challenge. It is primarily this reading that, throughout the rest of the paper, I attempt to refute in favour of an alternative non-Platonist one.

Let me already at this point announce what my main objection will be. Most, if not all, Platonizing interpretations, including and in particular Viljanen's, rely on the distinction between *essentia formalis* and *essentia actualis*. That Spinoza operates with such a dichotomy has, at this juncture, become so entrenched in scholarship that commentators do not question it and even tend to see it where it is in fact not present. We have already seen Garrett affirm that Spinoza 'invokes' that distinction in EVP21–22, while in reality Spinoza mentions neither formal nor actual essences in those propositions, but only speaks of actual existences and of essences *simpliciter*. For another good example, one can consult a recent paper by Robert Schnepf who affirms that 'anyone asking himself how to interpret the concept of essence, for instance, can be quickly referred to EIIP8, but also to the beginning of the third part of the *Ethics*, in particular to EIIIP7, where Spinoza speaks of *essentia actualis* in

and Y. Melamed, 'Spinoza's Metaphysics of Thought: Parallelism and the Multi-Faceted Structure of Ideas', *Philosophy and Phenomenological Research* 86:3 (2013): 636–83.

[43] This chapter, note 7.

[44] See Scribano, *Guida alla lettura dell'Etica di Spinoza*, p. 55.

contrast to *essentia formalis*.'[45] Spinoza, of course, makes no such distinction in EIIIP7, where the notion of formal essence is nowhere in sight! Moreover, in the context of a philosopher whose basic mindset is heavily influenced by Descartes, as Spinoza's indisputably is, opposing the terms 'actual' and 'formal' is quite surprising, since Descartes rather tends to identify them. For example, in *Meditation* III, Descartes explicitly employs the expression *realitas actualis sive formalis* and, in the *Principia philosophiae*, I, §17, he again uses the notions of 'actual reality' and 'formal reality' as synonymous.[46] While similarities to Descartes should always be considered with caution, this nonetheless suggests that the dichotomy between formal essences and actual essences is not as obvious as some commentators would like to think, but that, maybe, Spinoza never intended to oppose, or even juxtapose, those two terms. Overcoming this false dichotomy is the principal polemical aim of the following section.

## 1.2 Formal Essence

Let us first take a closer look at Spinoza's use of the notion of 'formal essence' and the precise contexts in which it occurs. The notion of formal essence first appears in the polemical context of EIP17S:

> If the intellect pertains to the divine nature, it will not be able to be (like our intellect) by nature either posterior to (as most would have it), or simultaneous with, the things understood, since God is prior in causality to all things (by P16C1). On the contrary, the truth and formal essence of things is what it is because it exists objectively in that way in God's intellect.

Spinoza here argues that if the divine intellect is part of the divine essence, this divine intellect cannot be like ours, since our intellect requires an object. For, on this hypothesis, the ideas in the divine intellect would not be true because they objectively express what formally exists, but be true simply in virtue of being conceived in the divine intellect which, as a component of the divine essence, would be prior to all things, including its own object. What we have here is a version of the theory of divine prescience and creation: An omniscient God first conceives the world and then creates that world according to that plan. Now, as already pointed out above, Spinoza himself does not endorse the premise of this argument. As is clear from EIP31, the intellect, even infinite, does *not* pertain to the divine nature for Spinoza. It is something that *follows* from that nature, as an immediate infinite mode. Consequently, he does

[45] R. Schnepf, 'The One Substance and Finite Things (1P16-28)', in *Spinoza's Ethics. A Collective Commentary*, edited by M. Hampe, U. Renz, and R. Schnepf (Leiden & Boston: Brill, 2011), 37–56, p. 56.

[46] For Descartes, see *Principles of Philosophy*, I, art. 17, CSM I, 199/AT VIIIA, 11: 'All the intricacy which is contained in the idea merely objectively... must be contained in its cause... not merely objectively or representatively, but in actual reality, either formally or eminently'; and *Metaphysical Meditations*, III, CSM II, 28/AT VII, 41: '... this cause does not transfer any of its actual or formal reality to my idea....' I am indebted to Oberto Marrama for these references.

not endorse any of the conclusions of the argument either. That the divine intellect is 'not like our intellect' is the consequence that he deems most absurd. For Spinoza, our intellect is *exactly* like God's because it is a part of it (EIIP11C). This is the basic premise of his exceptional epistemological optimism. But Spinoza does not, and this is what is most relevant to us here, endorse the second consequence either, namely that the formal essence of things is what it is because it exists objectively in God's intellect. Quite to the contrary, as Spinoza later writes in EIIP6C:

> [T]he formal being of things which are not modes of thinking does not follow from the divine nature because [God] has first known the things; rather the objects of ideas follow and are inferred from their attributes in the same way and by the same necessity as that with which we have shown ideas to follow from the attribute of thought.

What formal essences *then* are becomes a bit clearer in EIIP8:

> The ideas of singular things, or modes, that do not exist must be comprehended in God's infinite idea in the same way as the formal essences of the singular things, or modes, are contained in God's attributes.

Formal essences are what they are, because they are contained in God's attributes, not his intellect.[47] Consequently, what we find in the divine intellect is not formal essence, but what Spinoza terms *objective being*. The proposition is followed by a corollary according to which 'so long as singular things do not exist, except insofar as they are comprehended in God's attributes, their objective being, or ideas, do not exist except insofar as God's infinite idea exists' (EIIP8C). Moreover, according to Spinoza, EIIP8 is 'evident from the preceding one', i.e. EIIP7, according to which 'the order and connection of ideas is the same as the order and connection of things'. However, the truth of EIIP8 is not exactly evident from EIIP7 itself. What I think Spinoza has in mind is in fact the *corollary* of EIIP7, according to which 'whatever follows formally from God's infinite nature follows objectively in God from his idea in the same order and with the same connection.' From this it is indeed quite clear that whenever I have an adequate idea of a thing, or that the thing is objectively given in the intellect, there must be a corresponding formal essence of that thing in the relevant attribute.

From these various passages clustered around EIIP7C, EIIP8 and EIIP8C, emerges the following configuration of notions. First, a distinction between *formal essence* and *objective being*, also called *ideas*. Next, a corresponding distinction between *what*

[47] *Contra* Henri Krop who maintains that 'the formal essence is an essence in itself, but it exists independently of the things. It is uncreated, although, unlike God, it does not exist in itself (CM 1.2) and exists objectively—and eternally—in God's intellect (EIp17s). The sequence of formal essences corresponds with the eternal sequence of essences in God' (H. Krop, '*Essentia*' in *Continuum Handbook to Spinoza*, edited by W. van Bunge et al. (New York: Continuum, 2011), p. 211. We note the reference to Spinoza's polemical development in EIP17S, which, like in Martin, is taken in the affirmative rather than as part of a *reductio*. The faulty notion that formal essences are in the intellect of God stems from this misreading.

*follows formally from God's infinite nature* and *what follows objectively in God*. Spinoza clearly taps into the traditional Cartesian distinction between the *formal*, or *real*, and the *objective*, or *representative*.[48] Now, what follows formally from God's infinite nature is, according to EIP16, 'infinitely many things in infinitely many modes', that is to say, the totality of modes produced in all the attributes of God. What follows objectively in God, on the contrary, is 'everything which can fall under an infinite intellect', as Spinoza puts it in EIP16, that is to say, the totality of the ideas of all those modes as they are comprehended in the divine intellect. Hence, the content of the ideas in the divine intellect, insofar as such ideas are representations, is the objective being of the corresponding formal essences as they are contained in the attributes. As Spinoza puts it in the *Korte Verhandeling*, clearly prefiguring the *Ethics*' conception of the divine intellect as the immediate infinite mode of thought, 'the most immediate mode of the attribute we call thought has objectively in itself the formal essence of all things', and there is 'an infinite Idea, which contains in itself objectively the whole of Nature, as it is in itself', i.e. 'a creature created immediately by God, since it has in itself objectively the formal essences of all things' (KV, Appendix II, §3 and §5).

The notion of 'formal essences' appears yet again in EIIP40S2, where Spinoza explains that the third kind of knowledge, also called intuitive knowledge, 'proceeds from an adequate idea of the formal essence of certain attributes of God to the adequate knowledge of the essence of things.' Grasping what this third kind of knowledge consists in more precisely is not our concern at this point (I will return to the matter). The passage is however helpful in another respect. It confirms that the notion of 'formal essence' primarily serves its purpose in the context of determining how things relate to their representations in the intellect, i.e. how the formal essence of a mode in some attribute relates to the objective being of that same mode as represented in the intellect of God. But if that is the case, speaking of the formal essence is not speaking of some entity distinct from the essence *simpliciter*, but a way of distinguishing things themselves from their representations, or ideas. Insisting on the difference between the formal essence and the objective being of things is a way of making the point that we should not take the latter for the former, or that we should not mistake ideas in the intellect, even ideas in the divine intellect, for essences. Ideas, insofar as they are intellections, are ideas *about essences*, i.e. representations *of* what things are, while the essences themselves pertain not to the intellectual ideas, but to modes of the attributes of God, i.e. they are, ontologically speaking, the modes'

[48] Spinoza also frequently appeals to a distinction between the objective idea and the formal essence in the TIE: '[T]he idea is objectively in the same way as its object is really. So, if there were something in Nature that did not interact with other things, and if there were an objective essence of that thing which would have to agree completely with its formal essence, then that objective essence would not interact with other ideas...' (§41); 'We have shown that a true idea is simple, or composed of simple ideas...; and that its objective effects proceed in the soul according to the formal nature of its object' (§85); '...the intellect knows that things are formally as they are contained objectively in itself' (§107).

essence, i.e. *what* things are.[49] We thus return to the exact same point as the one Spinoza made in the polemical EIP17S where the notion of 'formal essence' first occurred. Contrary to those who think that the intellect of God pertains to his nature, and therefore confound the ideas of things (or objective being) with the essences of things (or formal essence), Spinoza stresses that formal essences are not ideas in the divine intellect, but that they pertain to the modes of the attributes, modes of which the ideas in the intellect of God are only representations. Hence, Spinoza does not introduce the notion of formal essence in order to posit a separate domain of essences. He uses the notion in order to distinguish the essences of things from their ideas, or representations. But at no point does he imply that the formal essence of a thing is different from the essence *simpliciter* of that thing, even less that it is distinct from actual essence. The formal essence of a thing is the formal being of the thing of which it is the essence, i.e. the essence insofar as it pertains to a mode that follows from an attribute of God, as opposed to the objective being of that thing, which is the representational content of an idea in the intellect of God.

## 1.3 Existence and Non-existence

As we have seen, Spinoza uses the notion of formal essence in contrast to the notion of objective being in the context of his attempt to provide a clear distinction between the essences and the ideas of things. There is however an important other context for the notion of formal essences that concerns his conception of non-existent things or the status of things that do not presently exist. Spinoza, we recall, writes the following in EIIP8:

> The ideas of singular things, or modes, that do not exist must be comprehended in God's infinite idea in the same way as the formal essences of the singular things, or modes, are contained in God's attributes.

As noted above, this proposition establishes that essences *are* not the ideas of things. But it also establishes that adequate ideas of singular things (i.e. the ideas that belong to the third kind of knowledge, or knowledge of singular essences) are ideas of essences, and moreover, that those ideas are comprehended in God even if the things of which those essences are essences do not exist. Earlier in the text, in EIP8S2, Spinoza already affirms about our comprehension that:

> we can have true ideas of modifications which do not exist; for though they do not actually exist outside the intellect, nevertheless their essences are comprehended in another in such a way that they can be conceived through it.

[49] From another point of view, such ideas do however have their own formal being, insofar as they are modes of the attribute of thought and, as such, have formal essences. As such, they can be the object of yet other ideas (ideas of ideas). The important thing to grasp is that ideas *qua* intellections are not essences.

Hence, adequate ideas about singular non-existent things are not ideas without *ideata*, but there is an intellectual correlate in virtue of which such ideas, when adequate, can also be said to be *true*. For an idea is only true to the extent that it corresponds to its *ideatum* (EIA6).[50] Formal essences are thus introduced in order to account for the fact that ideas about modes that do not exist are still ideas *about something*.

The problem, then, is to determine exactly the ontological status of such *ideata*, i.e. the kind of being, if any, we can bestow upon the formal essences of non-existing modes. Spinoza clearly affirms that such essences are somehow 'contained in God's attributes'. In that respect, however, they do not differ at all from the formal essences of *existing* modes. Indeed, for Spinoza, saying that a thing exists just means that the attributes of God are presently modified so that they produce that mode. Hence, the kind of existence that Spinoza ascribes to formal essences of non-existing things is not situated on any other level, or in any other ontological realm, than the kind of existence Spinoza also ascribes to the essences of existing things, i.e. things that presently exist in duration. I think this is the first, important point Spinoza wants to make. In EIIP8, he insists on his substance-mode ontology: there is nothing but modified substance, i.e. modes of substantial attributes. When Spinoza affirms that formal essences are contained in God's attributes, he is thus not only affirming that they are contained in the attributes, but also, and maybe more importantly, reiterating the claim that such non-existent essences are not contained *in anything else* than the very same attributes in which existent essences are also contained. Formal essences are contained in no heaven of forms; there is no transcendental divine intellect containing all ideas; no *regio possibilitatis* à la Leibniz, and so on. What we have in this passage, then, is primarily a forceful rejection of transcendentalist versions of Platonism, i.e. of any conception of an otherworldly realm of pure essences.[51]

Formal essences must somehow be really contained in the attributes alongside the actual existences but without actually existing. Should we then conclude that each attribute contains *two different modes* each expressing the same thing under that same attribute, the one expressing the formal essence and the other the actual existence of that thing? Any such doctrine, as argued in the previous section of this essay, contradicts the principles of Spinoza's parallelism. Two modes of the same attribute cannot express the same thing. But we cannot account for the reality of formal essences by positing a mysterious ontological domain of 'non-existing existence' beyond that of actual, existing existence, a distinct domain for the eternal 'being of essence', without embracing the Platonizing conception of an otherworldly realm of pure essences that Spinoza clearly argues against. Finally, as Martial Gueroult stresses, we should be particularly wary about attributing to Spinoza the quasi-Averroistic notion that essences disappear into the generality of the attribute

[50] See also EIID4Exp. and EIIP32D.

[51] See Gueroult, *Spinoza I: Dieu*, p. 331.

when the corresponding existences go out of duration. Formal essences are not contained in the attributes as mere potentialities, like statues in a block of marble.[52] They cannot be, since what is, is *in actu* for Spinoza.[53] Moreover, it would be ruinous for Spinoza's later conception of intuitive knowledge to suggest such a theory, since intuitive knowledge is exactly supposed to be knowledge about the essences of *singular* things, independently of their existence and non-existence (EVP36S). If the essences of non-existing modes dissolved into the generality of the attribute, intuitive knowledge of anything non-existing would be either impossible or objectless. And yet, we cannot follow Gueroult in concluding that 'essences thus have a reality in act that is distinct from the reality in act of their existence.'[54]

So what options do we have left? EIP8S2 suggests that the formal essences of non-existent modes are contained in the attributes in virtue of *something else* than that mode itself. He claims that 'their essences are comprehended in another [*in alio comprehentur*] in such a way that they can be conceived through it.' What could Spinoza possibly mean by that? By *in alio*, does Spinoza here simply mean God, or the attributes of God? In that case, without further specification, we seem to return to the unacceptable scenario where formal essences are contained in a merely general way in the attributes, like statues in marble. Referring the conception of non-existing modes exclusively to their comprehension in an attribute is far too general to account for their essence. For example, conceiving some body merely as extended teaches us little about what that body is, and absolutely nothing about what distinguishes it from other bodies. But this does not seem to be what Spinoza had in mind in EIP8S2, since he claims that conceiving a non-existing thing through its 'comprehension in another' *does* earn us knowledge of that non-existing thing's essence.

I think Spinoza's puzzling phrase in EIP8S2 must be understood in the context of the principle he formulates in EIP11D2, according to which 'for each thing there must be assigned a cause, *or* reason, both for its existence and for its nonexistence.' As has been remarked often enough, this principle of reason is distinguished from its famous Leibnizian counterpart, the principle of sufficient reason, by adding the additional requirement of providing reasons for non-existence, and not only for

[52] Stuart Hampshire, Edwin Curley, and Elhanan Yakira go in this direction in their readings of Spinoza's conception of the immortality of the mind, that is to say, in their understanding of what, for Spinoza, remains of the human mind—which for Spinoza is nothing but the idea of the body and a mode of the attribute of thought—when that mode no longer exists. See S. Hampshire, *Spinoza* (Harmondsworth: Penguin, 1951), p. 175; E. Curley, *Behind the Geometrical Method. A Reading of Spinoza's Ethics* (Princeton: Princeton University Press, 1988), pp. 84–6. For Yakira, non-existence, as opposed to actual existence or existence in act, is 'undifferentiated' ('Ideas of Nonexistent Modes', p. 159 and p. 169). On these questions, see also S. Nadler, *Spinoza's Heresy. Immortality and the Jewish Mind* (Oxford: Oxford University Press, 2001), pp. 105–6 (on Hampshire and Curley), and pp. 129–30 (for Nadler's own revised and more sophisticated version of this interpretation). For my reading, see M. Lærke, "Spinoza on the Eternity of the Mind", *Dialogue* 55:2 (2016): 265–86.

[53] See EIP36: 'Nothing exists from whose nature some effect does not follow.'

[54] See Gueroult, *Spinoza II: l'âme*, pp. 100–1.

existence.[55] This principle provides the ground for one of the defining features of Spinoza's modal philosophy, namely his rejection of *possibilia* in EIP33S.[56] For a conceivable thing to be non-existent is not to exist in some state of indeterminate possibility, but not to exist for causes or reasons that are just as determinate and real as the causes or reasons for which that which exists, exists. Both existence and non-existence must be accounted for by the internal causal relations among modes produced within the attribute.

Now, since for a thing to exist for Spinoza just means that the divine attributes are modified in such a way so as to produce this thing, one implication of the causal principle in EIP11D2 could seem to be that, just as existent things *are* contained in the attributes as existing in virtue of some external cause, non-existent things must be *not* contained in the attributes of substance as existing in virtue of a given external cause or reason. Such non-existent things are not just *absent* from existence, but positively *excluded* from existence by other, existent things. And when they are thus excluded from existence, this means that the unique substance is thus modified so as *not* to produce them. Those things simply do not exist. If, I suggest, Spinoza nonetheless holds that the formal essences of non-existing things *are* somehow contained in the attributes, it is because, even if non-existing things are *not* contained in the attributes *qua* existing, they are nonetheless contained in the attributes *qua* non-existing. They are modes presently excluded from existence by that which exists and thus contained in the attributes in a determinate way *qua* excluded from existing. It is in virtue of such determinate causes or reasons of non-existence that we can say that the formal essences of non-existing things are actually contained in the attribute *qua* non-existing.

This may sound rather esoteric, but there is a reasonably straightforward intuition behind the point. We must keep in mind the motivation of EIIP8, i.e. what it is Spinoza hopes to establish by means of that proposition. I think it is to ascertain that there can be not only adequate, but also *true* ideas about things that do not exist, i.e. that such ideas are truthfully 'comprehended in God's infinite intellect.' If we follow EIAx4 and TIE §92, Spinoza is committed to the view that an adequate definition of a given thing must be a *genetic* one. The adequate knowledge of a thing involves knowledge of its cause. Now, as is clear from EIID4, whether an idea is adequate or not, contrary to its truth, does not involve the consideration of the relation to the external object. I can form an adequate idea of a circle by conceiving the way in which

[55] See for example V. Carraud, *Causa sive ratio. La raison de la cause de Suarez à Leibniz* (Paris: Presses universitaires de France, 2002), pp. 326–9.

[56] Spinoza here denies that the notion of possibility has any kind of ontological relevance. Possibility cannot refer to a class of beings. This does not, however, mean that the notion possibility does not have epistemological relevance. Quite the contrary—in EIVD4, Spinoza himself defines possibility as a particular kind of ignorance, namely ignorance of whether the determining productive causes of a given thing are present or not. On this point, see also EIP33S, EIIP31C, and CM I, iii.

such a circle can be produced, e.g. by the rotation of a line around a point (TIE §96).[57] This genetic conception of the circle will be adequate whether such a circle exists or not. The *truth* of such an idea, however, is a different matter. For example, I can form an adequate idea of the ice-cream I ate yesterday by conceiving a genetic definition of that ice-cream, i.e. conceiving how it was generated by mixing cream, eggs, sugar, putting it in the freezer, and so on. When I have the whole recipe present to my mind, I have an adequate idea of that ice-cream. Is that idea then also *true*? Well, yes and no. It has all the 'intrinsic denominations of a true idea', as Spinoza puts it in EIID4. The problem is: I have gobbled up the ice-cream. So my adequate idea of the ice-cream no longer has an existing object. I ate it. Consequently, it seems, the idea no longer has the 'extrinsic denominations' of a true idea, which are that it should 'agree with its object' (EIAx6). And yet, in a sense, it *does* have an object, namely the ice-cream *qua* eaten, or *qua* a non-existing thing that is determinately expressed in the attributes in virtue of my existence as the cause of its disappearance. How? In order for my adequate idea about yesterday's ice-cream to be true, it must include also the notion that it is no more. This, however, cannot simply be conceived by *abstracting* from the causes of the thing, on pain of the idea becoming inadequate, since an adequate idea involves knowledge of the cause. In itself, the adequate idea of a thing necessarily posits the thing as existing exactly because that idea involves the idea of the cause that brings into existence the thing. Consequently, the only way in which a conception of yesterday's ice-cream can be said to be true is if that idea includes not only the idea of the causes of the ice-cream's past existence but *also* of the causes of its present non-existence, i.e. a conception of yesterday's ice-cream *qua* eaten by yours truly. Hence, in general terms, the idea of a non-existing thing can be said to be true, have a real extra-intellectual object, or be somehow contained in the divine attributes, in virtue of the really existing things that causally account for its present non-existence, thus allowing us to posit the thing as contained in the attributes *qua* non-existent.

Admittedly, Spinoza is not quite as explicit as I would like him to be in making this connection between the axiomatic requirement of causes of non-existence in EIP11D2 and the notion of 'modes that do not exist' in EIIP8, but I do not think it is particularly awkward. Moreover, I think Spinoza already hints at something like this in EIP8S when asserting about non-existing things that 'their essences are comprehended *in another* in such a way that *they can be conceived through it*' (my italics). The reality of non-existing things, their formal essences, can be conceived through the things that do presently exist in virtue of the fact that the latter exclude the former's existence. Non-existent things cannot, of course, be conceived *as existing* through the things that exclude their existence. But the latter nonetheless permits

[57] See Spinoza, Letter 9. This point is well-documented. See, for example, E. Curley, *Behind the Geometrical Method*, p. 111; and H. De Dijn, *Spinoza: The Way to Wisdom* (West Lafayette, Indiana: Purdue University Press 1996), pp. 155–7.

conceiving the essence of such things as *actually non-existing*. Yesterday's ice-cream haunts my present existence. Not only has it made me fatter, but its formal essence remains really outlined in the attributes of God *qua* non-existing by reason of *my* existence *qua* fatter. It is in virtue of such indirect reference that I can say that my adequate idea of a non-existing thing is also *true*, i.e. has an extra-intellectual object to conform to, or a real *ideatum* in the attributes of God.[58]

This leaves us with the following, final question. When, objectively, I grasp the essence of a thing, be it as existing or as non-existing, *what* is it that I grasp? What, exactly, does a formal essence *look like*? We need here to consider Spinoza's notion of 'form'. Indeed, content-wise, I think we take 'formal essence' to refer to the formal aspect of the essence, or the aspect of the essence that relates to the form of the thing. The close connection between Spinoza's notions of 'essence' and 'form' is clear from EIIP10, where Spinoza uses the notions as equivalent when stating that 'the being of substance does not pertain to the essence of man, or substance does not constitute the form of man.'[59] But what, then, is a 'form' for Spinoza? As one could expect from a Cartesian, when it comes to bodies, the form of a finite thing can be formulated in terms of mechanical properties of movement, shape, and proportion. Hence, in the so-called 'Physical Digression' following EIIP13, Spinoza defines an individual as a unique structure or relative configuration of constituent parts which, under the attribute of extension, is a certain proportion (*ratio*) of motion and rest (see the Definition following Lemma 3, and Lemmas 5–6). This unique structure is what he also calls the 'form' of the individual. So, the form of a given mode is the structure of the mode, i.e. the relative configuration of its constituent parts. The formal essence is the aspect of the essence of a thing that relates to this form.[60] This invariable and eternal form is, depending on the time and place, contained in the attributes as existent or as non-existent, but always in one way or in the other.

Hence, to sum up, the formal essence of a thing is the extra-intellectual correlate of the idea through which we conceive the thing. It refers to the specific relative configuration of parts that defines the form and the individuality of the thing. That form is *always* contained in the attributes, either as existing or as non-existing. When the thing exists, its form is contained in the attributes in virtue of the causes that

[58] One might here wonder how this approach deals with the question of *fictions*, such as when I 'imagine a winged horse' following Spinoza's example in EIIP49S. It should here be stressed that, for Spinoza, fictional ideas are not ideas about non-existing things; they are ideas about existing things or modes, namely those that Spinoza in EIIP17S calls 'images' (under the attribute of extension) or 'imaginations of the mind' (under the attribute of thought). Hence, the true object of fictional ideas, i.e. object of the idea God has of such fictions, is not a non-existing thing, but an existing modification of a singular mind imagining that fiction.

[59] We can also consider the Lemmas 4, 5, and 6 in the 'Physical Digression' following EIIP13, where Spinoza uses the notion of 'form' as quasi equivalent to that of 'nature' when describing the circumstances under which an individual will 'retain its nature, as before, without any change of its form.'

[60] See Nadler, *Spinoza's Heresy*, p. 111; and Nadler, 'Spinoza's Monism and the Reality of the Finite', p. 234.

brings it about; when not, the form is contained in the attributes in virtue of the causes that exclude it from existence. Contrary to what is asserted by the Platonizing readings, then, grasping the formal essence of the thing is not grasping an entity beyond or other than the thing itself (as Scribano and Schmaltz seemed to hold) or even grasping the thing itself as existent in an non-actual way that is different from actual existence (as Viljanen suggested). It is grasping the form of the thing itself, its specific structure, insofar as this thing is contained in the attributes either *qua* existent, through the causes that makes it exist, or *qua* non-existent, through the causes that exclude it from existence.[61] That form is objectively conceived by the intellect as an eternal truth and the form itself eternally contained in the attribute.[62]

## 1.4 Actual Essence

Let me now turn to the notion of 'actual essence'. It first appears in EIIIP7: 'The striving by which each thing strives to persevere in its being is nothing but the actual essence of the thing.' It reappears briefly in EIVP4D, according to which 'the power by which singular things . . . preserve their being is the power itself of God, or Nature, not insofar as it is infinite, but insofar as it can be explained through the man's actual essence.' Spinoza does not use the notion of 'actual essence' elsewhere. Much less does he define what he understands by it.

He does, however, make some interesting remarks about the term 'actual'. We can first consult some passages in the first part of the *Ethics*. In EIP33S, Spinoza joins 'all the Philosophers [he has] seen' in affirming that 'in God there is no potential intellect, but only an actual one.' In EIP31S, Spinoza has already warned that 'the reason why I speak here of actual intellect is not because I concede that there is any potential intellect, but because, wishing to avoid all confusion, I wanted to speak only of what we perceive as clearly as possible, i.e. of the intellection itself. We perceive nothing more clearly than that.' What Spinoza says here is that it makes little sense to add the qualification 'actual' for other than polemical reasons, that is to say, to stress a point that otherwise could be misunderstood by an interlocutor. In reality, there is no conceivable 'potential intellect' opposed to and/or distinct from the actual one. This is consistent with Spinoza's modal philosophy from which possibility and contingency have been banished as ontologically irrelevant. Things are only called contingent or possible 'because of a defect in our knowledge.' (EIP33S1; see also EIVD3–4). Qualifying a thing as actual adds nothing to the concept of that thing, but serves the purpose of directing our attention towards a feature already included in that concept, namely that, when adequate, it is always a concept of a thing *in actu*, and that a concept of a thing that is not a concept of a thing *in actu* is a defective one.

[61] For a similar point, see P.-F. Moreau, *Spinoza. L'expérience et l'éternité* (Paris: Presses universitaires de France, 1994), p. 501.

[62] See Nadler, *Spinoza's Heresy*, p. 111.

In a sense, it is simply redundant to add the qualification 'actual' (because it is inconceivable that anything would not be so), *except* in order to highlight a particular *aspect* of the thing qualified. In this case, in EIP31S, Spinoza wants to stress by means of the locution 'actual intellect' that he speaks of 'intellection itself', that is to say, the actual act of understanding, which we, in that very act, 'perceive as clearly as possible.' Adding 'actual' here simply serves the purpose of eliminating the notion that there is any such thing as a faculty of understanding distinct from the actual acts of understanding, thus already preparing the ground for the theory developed in EIIP48 and EIIP49D, according to which 'there is in the Mind no absolute faculty of understanding' but that 'the intellect is nothing apart from the singular... ideas themselves.'

Spinoza employs the adjective 'actual' when speaking of 'actual essence' in EIIIP7 in exactly the same way. That this is a correct interpretation is indicated by the ways in which Spinoza reformulates EIIIP7 when referring to it in later demonstrations. Hence, in EIVP22D, 'the striving to preserve itself is the very essence of a thing (by IIIP7)'; in EIVP25D, 'the striving by which each thing strives to persevere in its being is defined by the thing's essence alone', or again in EIVP26D, 'the striving to preserve itself is nothing but the essence of the thing itself (by IIIP7).' It is remarkable that Spinoza, when appealing to EIIIP7, *never* deems it necessary to reiterate the status of the 'essence' in question as an 'actual' one. Apparently, in all these reformulations, the actuality is just *implied*. Indeed, it seems to make little difference to Spinoza whether he speaks about the actual essence or the essence *simpliciter*. In light of what Spinoza said about actuality in the first part of the *Ethics*, however, this is by no means surprising. Conceptually, the 'actuality' of the essence adds absolutely nothing to the essence itself, but simply serves the purpose of highlighting an aspect of it, a feature already included in the essence as such, namely the fact that things are essentially powerful and that that power cannot be conceived as mere potentiality. In other words, the power of things is *given* in and by their essence, such as Spinoza also implicitly defines the actuality of essence in EIIIP7C: 'the power, or striving, by which [a thing] strives to persevere in its being, is nothing but the given, *or* actual [*datam sive actualem*], essence of the thing itself.' Consequently, when Spinoza speaks of the *conatus* as the 'actual essence' of things, this is not to imply that we should distinguish this actual essence from some other, not actual essence, be it the essence *simpliciter* or the formal essence. His aim is quite to the contrary to stress that, in the same way as all things display a form or characteristic *ratio*, which is the formal aspect of its essence, all things also display a distinctive power or *conatus*, the actual aspect of its essence, that makes that thing persevere in existence. All things have an actual aspect of power or *conatus* that need not be accounted for by anything over and above their essence.[63] This actuality of a thing's power, or *conatus*, pertains to the essence of the thing regardless of the thing's existence. This is

[63] See EIIIP4: 'No thing can be destroyed except through an external cause' (EIIIP4). Such external causes or reasons provide the explanatory framework for *determinate* existence in duration, as opposed to

not to say that things that do not exist have a *conatus* or that they persevere in existence even when they do not exist. Non-existent things clearly cannot persevere in an *existence* that they do not have. But that in no way prevents their *essence* from being actual rather than potential power. To say that the actual essence of the thing is power simply means that when we conceive adequately of a thing we conceive of it as something that exercises an actual power and not just a potential one. But this is true of the thing whether it exists or not.

Now, this interpretation of the notion of 'actual essence' encounters a potential difficulty in relation to EIP24C, where Spinoza affirms the following:

> God is not only the cause of things' beginning to exist, but also of their persevering in existence.... For—whether the things [produced] exist or not—so long as we attend to their essence, we shall find that it involves neither existence nor duration. So their essence can be the cause neither of their existence nor of their duration, but only God, to whose nature alone it pertains to exist.

A Platonizing reading might explain this passage as follows. By the 'essence' of things Spinoza refers to the *formal essences* which, although they do have eternal existence, have no power in themselves to enter durational existence. Formal essences only acquire a *conatus* or *actual essence* when brought into durational existence by God. This passage would then affirm that there are unactualized formal essences and that something must be added to such essences for them to enter into actual existence, namely an actual essence or power that is produced by God by an act of causation which is distinct from the act of causation through which God produces the formal essences.

I do not think, however, that this is what Spinoza had in mind when asserting that the essence of a thing produced by God 'involves neither existence nor duration.' By 'essence', Spinoza does indeed refer to the formal aspect of essence, i.e. the internal configuration of parts that constitutes its form. But he is not separating the essence of a thing from the existing thing of which it is the essence, the formal essence from the actual essence. Rather, his intention is to stress that, contrary to the formal aspect of a thing's essence, the actual aspect of a thing's essence, its 'persevering in existence', must be understood and explained by the fact that things are essentially modes and not substances or, as Spinoza puts it in the corollary of the following proposition, that 'particular things are nothing but affections of God's attributes, or modes by which God's attributes are expressed in a certain and determinate way' (EIP25C). If all existing things persevere in existence in virtue of their essence, or in virtue of the actual aspect of their essence, it is because their essence just *is* to be in God, or to be a mode of God, whose power is his essence (EIP34).[64] In other words, things that exist have actual

the *indeterminate* existence, or indefinite duration, which necessarily follows from the actual aspect of the essence (EIIIP8). On this point, see also Viljanen, 'Spinoza on Virtue and Eternity', pp. 260–1.

[64] See also EIIP45S: 'I am speaking, I say, of the very existence of singular things insofar as they are in God. For even if each one is determined by another singular thing to exist in a certain way, still the force by which each one perseveres in existing follows from the eternal necessity of God's nature.'

essence, or power to persevere in existence, in virtue of the fact that they are essentially modes of God whose essence it is to exist from his power alone. If existing things have power, it is because they are essentially 'chunks' or 'portions' of God's power.[65] As Spinoza puts it in EV30D, such essences 'are conceived... as real beings... insofar as through God's essence they involve existence.' But none of this implies that we should deprive the essences of non-existing things of their actuality, even less that we should distinguish powerless formal essences from actual powerful ones!

## 1.5 Two Types of Actuality

I have above outlined an aspectual reading of Spinoza's notions of formal essence, existence and non-existence, and actual essence, which at no point requires that we appeal to the distinct ontological level that Platonizing readings introduce. So where does all the confusion come from? I believe Spinoza himself is partly responsible. There is a tension in his use of the adjective 'actual'. For it qualifies not only the notion of *essence*, like in EIIIP7, but also the notion of *existence*. Hence, the notion that something 'actually exists' occurs quite frequently in the *Ethics*. What does 'actuality' mean in this context? Let me first note that Spinoza frequently seems to use the terms 'actual existence' and 'existence' *simpliciter* as equivalent. In §55 and §58 of the *Tractatus de intellectus emendatione*, Spinoza simply identifies 'actuality' and 'existence'.[66] Once again, as was also the case when speaking of essences, the adjective 'actual' does not seem to modify the concept of existence it qualifies. It appears to be strictly speaking redundant. So why does Spinoza use the adjective 'actual' to qualify existence anyway?

Considering the occurrences in the *Ethics*, the only apparent difference I find between 'existence' and 'actual existence' is that Spinoza adds 'actual' when speaking specifically about the way in which the mind perceives the thing in question, i.e. when he addresses the relation of representation that exists between the existing thing and the mind that has an idea of it. Thus, for example, in EIIP11, 'the first thing that constitutes the actual being of a human Mind is nothing *but the idea of* a singular thing which actually exists'; and in EIIP13: 'The *object of the idea* constituting the human mind is the Body, or a certain mode of extension which actually exists, and nothing else' (my italics). But if that is the role of the adjective 'actual' in the context of existence, then it only serves to express *a way of perceiving* a thing, namely as unaccompanied by any *other* perception excluding its *present* existence. In other words, it is *perceived as enduring at present* in existence. So, 'actuality' says less about

[65] See Viljanen, *Spinoza's Geometry of Power*, pp. 74–5.

[66] See TIE §55: 'The same difference that exists between the essence of one thing and the essence of another also exists between the actuality or existence [*actualitatem aut existentiam*] of the one thing and the actuality or existence of the other [*actualitatem aut existentiam*]' and TIE §58: 'Let us pass now to fictions that concern either essences alone or essences together with some actuality *or* existence [*actualitate sive existentiam*].'

the concept of existence than it says about how the mind *perceives* this existence, namely as currently *present* or *not present* to it. That this is indeed the role of the adjective 'actual' when applied to existence is particularly clear from EIIP17: 'If the human Body is affected with a mode that involves the nature of an external body, the human Mind will regard the same external body as actually existing, or as present to it, until the Body is affected by an affect that excludes its existence or presence of that body.'[67]

Thus, in sum, Spinoza uses the term 'actual' to qualify both essence and existence. In both cases, however, actuality adds nothing new to what it qualifies but only serves the purpose of highlighting a specific aspect already implied in the notions of essence and existence. There is no essence that is not actual and there is no existence that is not actual. That, however, does not imply that the essence of things is always actual in the same sense as the existence of things is always actual. In fact, essences and existences are actual in two distinct ways. Essence is always actual in the sense that it is always an essence *in actu*. A thing is what it is, i.e. has the essence that it has, actually and not just potentially. Existence is always actual in the sense that it is always *perceived as present* by the mind. Consequently, the *essence* of a thing can perfectly well be actual even though the thing does not actually *exist*, since 'actual' does not mean the same thing in the two contexts.

This ambiguity in Spinoza's notion of actuality is, I think, irreducible from a strict textual point of view. But the difficulty is terminological rather than conceptual. It could have been resolved by giving a different name to one of the two kinds of actuality. Moreover, it does not become manifestly problematic in the global deduction in the *Ethics* before Spinoza needs to think seriously about how singular essences can be perceived by the mind *sub specie aeternitatis*, i.e. when Spinoza has to formulate more precisely how the third kind of knowledge, intuitive knowledge, allows us to perceive, or make present to the mind, the essence of a singular thing independently of the durational existence of that thing. This only happens in part V. In EVP29, Spinoza sets out to demonstrate that:

> Whatever the mind understands under a species of eternity, it understands not from the fact that it conceives the Body's present actual existence, but from the fact that it conceives the Body's essence under a species of eternity.

[67] The pattern is similar for the remaining occurrences. See EIIP19D, 'the ideas of affections of the Body are in God insofar as he constitutes the nature of the human Mind, *or* the *human Mind perceives* the same affections (by P12), and consequently (by P16) the human body itself, as actually existing (by P17)'; EIIP47D: 'The human Mind has ideas from which *it perceives itself*, its own body, and external bodies as actually existing'; EIIIP28D: '[W]e strive, as far as we can, to *regard it* as present, *or* as actually existing'; EIIIP47S: '... as often as we recollect a thing—even though it does not actually exist—we still *regard it* as present...'; 'The *Mind* neither expresses the actual existence of its Body, nor conceives the Body's affections as actual, except while the Body endures (by IIP8C); consequently (by IIP26), *it conceives* no body as actually existing except while its body endures'; EVP22D: '*The mind* neither *expresses* the actual existence of its Body, nor conceives the Body's affections as actual, except while the Body endures'; EVP23S: '*Our mind*, therefore, can be said to endure... only insofar as *it* involves the actual existence of the body...'. (all italics mine).

Now, 'conceiving the Body's present actual existence' is clearly a question of conceiving the body as actual in the existential sense, i.e. as presently conceived by the mind as existing. This actual existence is durational: 'The Mind neither expresses the actual existence of its Body, nor conceives the Body's affections as actual, except while the Body endures (by IIP8C); consequently (by IIP26), it conceives no body as actually existing except while its body endures' (EVP21D). On the contrary, 'conceiving the Body's essence under a species of eternity' amounts to grasping the formal essence of the thing, or the formal aspect of the thing.[68] In other words, this is about grasping the essences of things as something which is formally, i.e. really, contained in the attributes of God, regardless of whether the things exist or not: 'To conceive things under a species of eternity... is to conceive things insofar as they are conceived through God's essence, as real beings, or insofar as through God's essence they involve existence' (EVP30D). Spinoza stresses once again that this essence is a 'real being', i.e. actually and not just potentially contained in the attributes, even when the thing of which it is the essence does not exist, hence that it is essentially actual. Otherwise, the knowledge we have of those essences would not be knowledge of the essence of *singular* things, contrary to what Spinoza explicitly asserts in EVP36S.[69] We would not perceive of them in a precise and determined way, but only in a general way, like statues in a block of marble. Consequently, in EVP29, the two meanings of actuality clash, since Spinoza must hold that intuitive knowledge involves grasping the actuality of a thing's essence regardless of the actual existence of the thing. This is why, in EVP29S, Spinoza suddenly feels the urge to explicitly distinguish between two types of actuality:

> We conceive of things as actual in two ways: either insofar as we conceive them to exist in relation to a certain time or place, or insofar as we conceive them to be contained in God and to follow from the necessity of the divine nature.

There is undeniably an element of deductive clumsiness in this late scholium distinguishing two types of actuality because it not only invites us, but positively obliges us to retrospectively reevaluate all previous occurrences of the qualification 'actual' in Spinoza's treatise in light of this distinction. I think Spinoza would have been better off had he used different terms for the two kinds of actuality from the beginning in order to avoid confusion, instead of trying to patch up this

[68] Hence, I do not disagree with the Platonizing readings that the essence the mind conceives *sub specie aeternitatis* in intuitive knowledge is the *formal essence*, although I would stress that this should be understood as the *formal aspect* of the essence. As Viljanen points out ('Spinoza on Virtue and Eternity', p. 265), that intuitive knowledge is concerned specifically with formal essence is also suggested by an addition to EIIP40S in the *Nagelate Schriften*, i.e. the Dutch version of the *Opera posthuma* established by Spinoza's friends, according to which intuitive knowledge is 'adequate knowledge of the [NS: formal] essence of things'. However, such additions should, as Viljanen also notes, not be emphasized too strongly, as their authenticity is hard to determine.

[69] See EVP36S: '[T]he knowledge of singular things I have called intuitive, or knowledge of the third kind....'

terminological problem as late as in EVP29S. The scholium does, however, also serve as confirmation of my previous analysis, exactly because we have already seen the tension between these 'two ways' in play in Spinoza's use of the adjective 'actual'. We have already felt the ambiguity at work in the text. So, as unsettling as this distinction between two types of actuality in EVP29S is, at least on my reading we know exactly why Spinoza feels the deductive pressure to introduce it: It is in order to distinguish between the essential actuality (which is indifferent to existence in duration) and existential actuality (which is not indifferent to existence in duration). The whole previous deductive system pushes him to introduce it, so he does. In EVP29S, Spinoza specifically introduces the distinction in order to account for the fact that things, no matter whether they exist at present in duration or not, always exist formally or really insofar as their formal essences are contained in the divine attributes. Even when not actually existing, or more precisely, when they are actually non-existing, things are still actual in the sense that pertains specifically to essences: they are not mere potentialities, but really contained formally in the attributes of God, either through the causes of their existence or through the causes of their non-existence. And when we conceive the formal essences in this way, as contained in the attributes of God in a determinate way, regardless of the present existence of the things of which they are the essences, we conceive both the actual and the formal aspect of those essences. This is the way we conceive essences when we conceive them intuitively.

## 1.6 Conclusion (and Tackling the Definition of Eternity)

How can we sum up the preceding analyses? According to the *Ethics*, there is nothing but a unique modified substance the infinitely many modes of which are expressed in infinitely many attributes. Spinoza's ontology has only one, single level: existence as such or 'existence itself' as Spinoza calls it, is a univocal term. All that exists is a unique, modified substance.[70] Platonizing readings of Spinoza's theory of essence are not in conformity with this basic tenet of Spinoza's ontology when they consider formal and actual essences as situated on two different ontological levels, assimilating them to the *esse essentiae* and *esse existentiae* of the scholastics. Such readings introduce irresolvable tensions in Spinoza's philosophy between the framework of substance and mode and the framework of existence and essence. My alternative, aspectual analysis of Spinoza's theory of essence and existence is, on the contrary, in conformity with the requirements of Spinoza's ontology of a unique, modified substance.

[70] Contra Rivaud, *Les notions d'essence et d'existence*, p. 80.

Much of the Platonizing confusion about Spinoza's conception of essences stems from the fact that, in Spinoza, the term 'actuality' applies differently to essences and to existences. In particular, this contributes substantially to the three faulty notions that formal and actual essences are distinct entities, that we can conceive of non-existing things as unactualized formal essences, and that the existence of things comes about as the actualization of their formal essences. I have shown that, once this terminological problem regarding 'actuality' has been addressed and we construct the notions of formal essence and actual essence while paying attention to the distinct contexts of Spinoza's use of them, it is perfectly possible to provide an account of his conception of essences that conforms to a single-level ontology.

On my interpretation, all things have an essence that has both a formal and an actual aspect to it. The formal aspect is the defining *form* of the thing—a form that is contained in the attributes in a determined and not just general way no matter whether the thing exists or not. The actual aspect is its defining *power* or *conatus* which defines a thing by the power it actually exerts rather than by some (for Spinoza illusory) potential that it might exert. Again, such actuality is a feature of the thing's essence whether the thing exists or not. Consequently, non-existing things are not unactualized formal essences and coming into existence is not what bestows actuality on *essences*. What non-existent modes do not have is actual *existence*, they are actually non-existent. Whether a thing exists or not, its *essence* thus has both a formal and an actual aspect. Therefore, we should *not* situate the formal essence and the actual essence of a thing on different ontological levels. These notions refer to two different aspects of one and the same essence. Moreover, those two aspects of essence are relevant to Spinoza in completely different contexts. They should not be opposed or even juxtaposed.

(1) Spinoza opposes 'formal essence' to 'objective being' as part of an argument against the notion that the essences of things are ideas contained in the divine intellect. For Spinoza, the essences of things are rather contained in the attributes and constitute the extra-intellectual, i.e. formal or real, *ideata* of the divine ideas.[71] Moreover, Spinoza appeals to the notion of formal essence in order to highlight the relation between the essence and the form of a thing. The formal essence is the aspect of the essence that relates to its form, or relative configuration of parts. Contrary to an Averroistic position, according to which non-existing essences are dissolved into the generality of the divine attributes, he maintains that such formal essences are really contained in the attributes of God, even when the things of which they are the essences do not

[71] To avoid all misunderstanding, I should here stress that the attributes containing the extra-intellectual correlate of such ideas include the attribute of thought. The attribute of thought is, for Spinoza, external to the intellect in the sense that ideas, qua contained in the intellect, are representations, whereas ideas, qua modes of the attribute of thought, are not. As such, a mode of the attribute of thought does not understand anything. It just is.

exist, since their form is contained in a determinate manner in the existing things that provide the precise cause or reason for their non-existence.

(2) Next, in a completely different context, Spinoza speaks of 'actual essence' when stressing that a thing is essentially a mode of a God whose essence is power, and hence it is of the essence of a thing to have such power, *conatus* or persevering in existence. Now, the actuality of essence refers to an aspect of a thing's nature, an aspect of *what* it is, and not to the fact *that* it is, if it is. If Spinoza qualifies essence as 'actual', it is thus not because there are any conceivable non-actual essences or that essences *simpliciter* require actualization in order to have power. He does this to stress that things, when they exist, essentially are the power that they actually exert rather than some (illusory) potential that they may or may not exert. It is in this sense that the *conatus* is 'given or actual' (EIIIP7D). This kind of 'givenness' pertains to the thing's essence no matter whether the thing exists or not. The essences of non-existing things are as actual as those of the existing ones, i.e. *what* they are is something that they are actually, and not potentially.

This leaves us with a final difficulty regarding Spinoza's definition of eternity that I will try to resolve as a kind of conclusion. According to the analysis developed above, existence, or being, is a univocal term in Spinoza. There is only one kind of existence, namely modified substance. Consequently, whether we ascribe eternal or durational existence to something, the property existence as such is always ascribed in the same sense. Eternal existence and durational existence cannot be two different kinds of existence, but must rather be one and the same existence *as* durational and *as* eternal, two distinct aspects of a same unique kind of existence. This interpretation, however, could appear to blatantly contradict Spinoza's definition of eternity, along with its explication:

> By eternity I understand existence itself, insofar as it is conceived to follow necessarily from the definition of an eternal thing. / Exp. For such existence, like the essence of the thing, is conceived as an eternal truth, and on that account cannot be explained by duration or time, even if the duration is conceived to be without beginning or end. (EID8&Exp.)

Does this not clearly indicate, as Tad Schmaltz suggested, that we should conceive of two forms of existence, eternal existence of essences and durational existence of existences, or a being of essences and a being of existences, to follow the scholastic distinction from the CM, and that those two kinds of existence/being cannot be explained in terms of each other? Should we not conclude that what Spinoza here calls 'existence itself', or 'eternity', and which he ascribes to essences, is distinct from durational determinate existence, and thus that he here plainly affirms the distinction between two kinds of existence that, above, I have much laboured to demolish?[72]

[72] Pierre-François Moreau correctly objects that Spinoza does not affirm in EID8Exp that the essences of things are eternal, but only that such essences are conceived as eternal truths, and suggests on this basis

I think not. If we consider EID8 more closely, it appears that Spinoza does not identify eternity with 'existence itself', but rather with 'existence itself, insofar as it follows from the definition of an eternal thing.' Consequently, it is not excluded that durational existence, or determinate existence, is also just 'existence itself', but insofar as it follows from some other kind of definition. Indeed, I think the appropriate dichotomy to establish here is not between *(eternal) existence itself* and *(durational) determinate existence*, but rather between *existence itself, insofar as it follows from the definition of an eternal thing*, on the one hand, and, on the other, *existence itself, insofar as it follows from the definition of a determinate thing*. This is, grammatically speaking, a no less natural reading of the passage than the Platonizing one, and in line with Spinoza's single-level ontology. In both cases, eternal and durational existence, we are in fact talking of one and the same 'existence itself', although it appears under two distinct aspects, eternity or duration, following whether existence is conceived indeterminately as a property of substance, or whether it is conceived determinately as a property of the modes of substance. The difference between eternity and duration, then, does not map on to two levels of being, *esse essentiae* and *esse existentiae*, but rather relates to Spinoza's distinction between *natura naturans* and *natura naturata* (EIP29S). When Spinoza affirms that we conceive essences as eternal truths, it amounts to the following: When we consider a thing's *essence*, we conceive the aspect of that thing that relates it directly to the *natura naturans*, or to the essence of God. This is the sort of conception that Spinoza calls 'intuitive knowledge', or the third kind of knowledge, which 'proceeds from an adequate idea of the formal essence of certain attributes of God to the adequate knowledge of the essence of things' (EIIP40S). By contrast, when we consider the *existence* of a thing, we conceive the aspect of the thing that relates it to the *natura naturata*, that is to say, we consider it as a mode related to other modes, that either exists or does not exist following whether those other modes cause the first mode's existence or non-existence. In short, we conceive the aspect of the thing that is merely durational.

## Bibliography

Alquié, F., *Le Rationalisme de Spinoza* (Paris: Presses universitaires de France, 1981).

Bennett, J., *A Study of Spinoza's Ethics* (Indianapolis: Hackett, 1984).

that the essences of things may always be eternal truths without however the essences being themselves eternal (see Moreau, *Spinoza. L'expérience et l'éternité*, pp. 511–13). I have difficulties accepting the second point. The truth of an idea consists in the fact that it agrees with its object (EIA6). Consequently, being conceived as an eternal truth implies being conceived as having an object to agree with, eternally. It could seem, then, that when Spinoza affirms that essences are conceived as eternal truths, this also amounts to affirming that the essences thus conceived are eternal. On this point, see also the passage from the CM already quoted above according to which Spinoza, in a certain sense, 'agree[s] with those who say that the essences of things are eternal' (CM I, ii).

Busse, L., 'Über die Bedeutung der Begriffe "essentia" und "existentia" bei Spinoza', *Vierteljahrschrift für wissenschaftliche Philosophie* 10 (1886): 283–306.

Carraud, V., *Causa sive ratio. La raison de la cause de Suarèz à Leibniz* (Paris: Presses universitaires de France, 2002).

Curley, E., *Behind the Geometrical Method. A Reading of Spinoza's Ethics* (Princeton: Princeton UP, 1988).

De Dijn, H., *Spinoza: The Way to Wisdom* (West Lafayette, Indiana: Purdue University Press, 1996).

Delahunty, R. J., *Spinoza* (London: Routledge & Kegan Paul, 1985).

Deleuze, G., *Spinoza et le problème de l'expression* (Paris: Minuit, 1968).

Donagan, A., *Spinoza* (Chicago: Chicago University Press, 1988).

Garrett, D., 'Spinoza on the Essence of the Human Body and the Part of the Mind that is Eternal', in *Cambridge Companion to Spinoza's Ethics*, edited by O. Koistinen (Cambridge: Cambridge University Press, 2009), 284–302.

Gueroult, M., *Spinoza I: Dieu* (Paris: Aubier-Montaigne, 1968).

Gueroult, M., *Spinoza II: l'âme* (Paris: Aubier-Montaigne, 1974).

Hampshire, S., *Spinoza* (Harmondsworth: Penguin, 1951).

Jarrett, C., 'Spinoza's Distinction between Essence and Existence', *Iyyun: The Jerusalem Philosophical Quarterly* 50 (2001): 245–52.

Kenny, A., 'The Cartesian Circle and the Eternal Truths', *The Journal of Philosophy* 67:19 (1970): 685–700.

Klein, J. R., '"Something of it remains": Spinoza and Gersonides on intellectual eternity', in *Spinoza and Medieval Jewish Philosophy*, edited by S. Nadler (Cambridge: Cambridge University Press, 2014), 301–46.

Koyré, A., 'Le Chien, constellation céleste, et le chien, animal aboyant', in *Révue de métaphysique et de morale* 55:1 (1950): 50–9.

Krop, H., '*Essentia*', *Continuum Handbook to Spinoza*, edited by W. van Bunge, H. Krop, P. Steenbakkers, and J. van de Ven (New York: Continuum, 2011), 210–11.

Lærke, M., *Leibniz lecteur de Spinoza. La genèse d'une opposition complexe* (Paris: Honoré Champion, 2008).

Lærke, M., 'Spinoza's Monism? What Monism?', in *Spinoza on Monism*, edited by P. Goff (New York: Palgrave Macmillan, 2012), 244–61.

Lærke, M., 'Spinoza and the Cosmological Argument According to Letter 12', *British Journal for the History of Philosophy* 21:1 (2013): 57–77.

Lærke, M., 'Leibniz on Spinoza's *Tractatus de intellectus emendatione*', in *The Young Spinoza. A Metaphysician in the Making*, edited by Y. Melamed (New York: Oxford University Press, 2015), 106–20.

Lærke, M., "Spinoza on the Eternity of the Mind", *Dialogue* 55:2 (2016): 265–86.

Marrama, O., 'The Dog that is a Heavenly Constellation and the Dog that is a Barking Animal by Alexandre Koyré,' in *The Leibniz Review* 24 (2014): 95–108.

Martin, C. P., 'The Framework of Essences in Spinoza's *Ethics*', *British Journal for the History of Philosophy* 16:3 (2008): 489–509.

Matson, W., 'Body, Essence and Mind Eternity in Spinoza', in *Spinoza: Issues and Directions*, edited by E. Curley and P.-F. Moreau (Leiden: E. J. Brill, 1990), 82–95.

Melamed, Y., 'Spinoza's Metaphysics of Thought: Parallelism and the Multi-Faceted Structure of Ideas', *Philosophy and Phenomenological Research* 86:3 (2013): 636–83.

Melamed, Y., *Spinoza's Metaphysics: Substance and Thought* (Oxford: Oxford University Press, 2013).

Melamed, Y., 'The Building Blocks of Spinoza's Metaphysics: Substance, Attributes, and Modes', in *Oxford Handbook to Spinoza*, edited by M. Della Rocca (Oxford: Oxford University Press, forthcoming; preprint available online on *oxfordhandbooks.com*. DOI: 10.1093/oxfordhb/9780195335828.013).

Moreau, P.-F., *Spinoza. L'expérience et l'éternité* (Paris: Presses universitaires de France, 1994).

Nadler, S., *Spinoza's Heresy. Immortality and the Jewish Mind* (Oxford: Oxford University Press, 2001).

Nadler, S., 'Spinoza's Monism and the Reality of the Finite', in *Spinoza on Monism*, edited by P. Goff (New York: Palgrave Macmillan, 2012), 223–43.

Rivaud, A., *Les notions d'essence et d'existence dans la philosophie de Spinoza* (Paris: Félix Alcan, 1905).

Schmaltz, T., 'Spinoza's Mediate Infinite Mode', *Journal of the History of Philosophy* 35:2 (1997): 199–235.

Schmaltz, T., 'Spinoza on Eternity and Duration: The 1663 Connection', in *The Young Spinoza. A Metaphysician in the Making*, edited by Y. Melamed (New York: Oxford UP, 2015), 205–20.

Schnepf, R., 'The One Substance and Finite Things (1P16-28)', in *Spinoza's* Ethics. *A Collective Commentary*, edited by M. Hampe, U. Renz, and R. Schnepf (Leiden & Boston: Brill, 2011), 37–56.

Scribano, E., *Guida alla lettura dell'Etica di Spinoza* (Rome-Bari: Editori Laterza, 2008).

Spinoza, B., *Opera*, edited by C. Gebhardt, vols I–IV (Heidelberg: Carl Winter Verlag, 1925).

Spinoza, B., *Opere*, edited by F. Mignini and O. Proietti. (Milano: Arnoldo Mondadori Editore, 2007).

Spinoza, B., *The Collected Works of Spinoza*, edited by E. Curley, vol. I (Princeton, New Jersey: Princeton University Press, 1985).

Viljanen, V., *Spinoza's Geometry of Power* (Cambridge: Cambridge University Press, 2011).

Viljanen, V., 'Spinoza on Virtue and Eternity', in *Essays on Spinoza's Ethical Theory*, edited by M. J. Kisner and A. Youpa (New York: Oxford University Press, 2014), 258–72.

Ward, T. M., 'Spinoza on the Essences of Modes', *British Journal for the History of Philosophy* 19:1 (2011): 19–46.

Yakira, E., 'Ideas of Nonexistent Modes: Ethics II Proposition 8, its Corollary and Scholium', in *Spinoza on Knowledge and the Human Mind*, edited by Y. Yovel and G. Segal (Brill: Leiden, 1994), 159–69.

# 2

# Wolff's Close Shave with Fatalism

*Stephan Leuenberger*

On 13 November 1723, a cabinet order issued by Friedrich Wilhelm I, the 'soldier king', gave Christian Wolff 48 hours to leave Prussia, on penalty of hanging. Wolff's view would not allow the punishment of army deserters, the king had been told. In the eyes of Wolff's detractors, that was a consequence of what they took to be his commitment to fatalism—the view that everything is necessary. Deserting soldiers had no other option, and could not be held responsible.[1,2]

Wolff had been a professor at the University of Halle for seventeen years. His expulsion became a *cause célèbre* in Europe, generating a voluminous polemical literature. Having already been influential in German universities through a series of pioneering textbooks in the vernacular, he now became a hero of the Enlightenment movement. Soon he took up a professorship in Marburg, and started writing a series of textbooks in Latin. In his own written contributions to the controversy, Wolff did not appeal to academic freedom. Indeed, he acknowledged that the freedom to philosophize had to be constrained when it conflicted with the interests of religion, of virtue, and of the state.[3] Rather, he was adamant that the attribution of fatalism—as well as a number of other accusations—was wrong.

In this paper, I shall investigate Wolff's accounts of possibility and necessity, and ask whether they are compatible with the denial of fatalism. I shall leave open the more general question whether any of Wolff's doctrine commits him to fatalism.[4]

[1] My thanks are due to the audience at the conference 'The Actual and the Possible', especially Jessica Leech and Sarah Rossiter, and to Matteo Favaretti Camposampiero, Mark Sinclair, and a referee for this volume for comments on a draft.

[2] For a detailed account of what happened, see C. Hinrichs, *Preussentum und Pietismus* (Göttingen: Vandenhoeck & Ruprecht, 1971), pp. 388–441.

[3] Wolff, *Ausführliche Nachricht von seinen eigenen Schrifften, die er in deutscher Sprache heraus gegeben* (Frankfurt, 2nd edition, 1733; reprinted Hildesheim: Georg Olms, 1973), §42.

[4] In particular, I shall not discuss two familiar chestnuts. First, whether the principle of sufficient reason, which has a prominent place in Wolff's philosophy, is compatible with contingency in the world. (Wolff denied this; see Wolff, *Der Vernünftige Gedancken von Gott, der Welt und der Seele des Menschen, auch allen Dingen überhaupt, anderer Theil, bestehend aus ausführlichen Anmerckungen* (Frankfurt, 4th edition, 1740; reprinted Hildesheim: Georg Olms, 1983), §5, for example.) Second, I shall not discuss whether the necessary existence of an omnipotent, omniscient and omnibenevolent God, which Wolff claims to prove

An interpreter of Wolff's views of modality is not short of relevant textual material. Wolff commented on the topic in numerous places in his vast corpus. The most detailed and systematic discussion is to be found in his *Ontology* (1736). In this book of 696 pages in total, the chapter 'Of the possible and impossible' runs to 25 pages, and the chapter 'Of the necessary and contingent' to 37. There is also much relevant discussion elsewhere in the book. Clearly, the topic of modality was important for Wolff, and not merely because he became embroiled in the controversy over fatalism. Modal concepts are central to his philosophy, playing a role in his explications of the concepts of being and existence as well as of philosophy itself. He defines the latter as the 'science of the possibles, in so far as they can be'.[5] A being is something that can exist (§134)[6]—whether it actually exists or not—while a non-being is something that cannot (§137). Finally, existence, or actuality, is defined as the 'supplement of possibility' (§174). A concept so close to the foundations of the definitional edifice needs proper elucidation.

My interpretation will draw mainly on Wolff's texts, and will give relatively little attention to the relationship of Wolff's ideas to those of other philosophers.[7] In particular, I shall not discuss to what extent Wolff's view can be traced back to Leibniz, who also struggled to find room for contingency in his system. In philosophical historiography, Wolff is often disparaged as a mere popularizer of the views of the older thinker. There is no denying that much of his system is Leibnizian in spirit. But there are three reasons for discussing Wolff's views in their own right. First, Wolff certainly did not see himself as a disciple, and was fond of emphasizing that he is only concerned with the truth of a view, and not with whether Leibniz or anyone else held it.[8] Secondly, Leibniz's influence is very hard to assess in questions of detail. Many of his writings, in particular most of those dealing with logical questions, were not available to Wolff. Thirdly, there are extensive scholarly debates about almost every aspect of Leibniz's views on modality, and there is no space to address those in an essay mainly devoted to another philosopher.

The plan of the chapter is as follows. In section 2.1, I introduce Wolff's definitions of the central modal concepts and raise some interpretive questions. In section 2.2, I present evidence that Wolff's account does commit him to fatalism, despite his

(see §21 of Wolff, *Theologia naturalis*, vol. II, Frankfurt and Leipzig, 1737) leaves room for contingency. It is arguably in the spirit of Wolff's philosophy to separate questions about modality as such from questions about the modal consequences of certain views about God. The former belong to ontology, the latter to natural theology. There is no entry for 'God' in the index of the *Ontology*.

[5] Wolff, *Philosophia rationalis sive logica. Praemittitur Discursus Praeliminaris De Philosophia in Genere* (Verona, 3rd edition, 1735), §27.

[6] Unless I indicate otherwise, my citations by paragraph number refer to Wolff, *Philosophia prima sive ontologia* (Frankfurt and Leipzig, 2nd edition, 1736; reprinted Hildesheim: Georg Olms, 1962).

[7] For an account of Wolff's views of modality that puts them in a larger historical perspective, see chapter 5 of Pape, *Tradition und Transformation der Modalität* (Hamburg: Meiner, 1966).

[8] See Wolff, *Ausführliche Nachricht von seinen eigenen Schriften, die er in deutscher Sprache heraus gegeben*, §72, as well as the end of the preface to that work, for example.

protestations. His distinction between absolute and hypothetical necessity, presented in section 2.3, is of only limited help, as I shall argue in section 2.4. In section 2.5, I propose to add another twist to the interpretation of the account, which allows him to avoid fatalism—but only just.

## 2.1 Wolff's Definitions of Modal Terms

The *Ontology* deploys the geometrical method, if somewhat loosely. Chapters are divided into numbered paragraphs—967 in all in the whole book. Within each paragraph, a definition or theorem is stated in italics. In the case of theorems, a proof—more or less rigorous as the case may be—is supplied. Indented text, in smaller font, adds examples and informal commentary. In contrast to Spinoza's procedure in the *Ethics*, Wolff does not explicitly label bits of text as theorems, definitions, proofs, or the like.[9]

The discussion of modality starts with a definition of impossibility: 'What involves a contradiction is called *impossible*' (§79).[10] Accordingly, what is possible is defined as what does not involve a contradiction, or is not impossible (§85). We might expect that what involves a contradiction is a proposition. But the two examples of impossibilities he gives are not of that category: a figure enclosed within two straight lines (*bilineum rectilineum*), and an iron wood (*lignum ferreum*). As an example of something possible, he offers an equilateral triangle (§85).

Given that these are not propositions, what is their ontological status? The *Ontology* is not helpful in this respect. In the *Latin Logic*, which precedes the *Ontology* by two years and is extensively referenced in it, it is concepts (*notiones*) that are said to be possible or impossible, depending on whether they involve a contradiction or not.[11] But while concepts and their properties are a suitable subject matter for a treatise on logic, one would expect them to figure less prominently in a work on metaphysics. Moreover, the idea that concepts are possible and impossible is hard to reconcile with the claims (§§102–3) that the possible does, and the impossible does not, answer to a concept. We can allow, of course, that a concept answers to another one. But it would be odd to allow such second-order concepts, but insist that none apply to contradiction-involving concepts. It is tempting to take Wolff's iron wood and figure enclosed within two straight lines to be propositional functions, but this would clearly be anachronistic. Normally there is no harm in calling them simply 'things'.

[9] He justifies his practice of omitting such labels in philosophical works in the Preface of *Vernünftige Gedancken von der Menschen Thun und Lassen* (Frankfurt and Leipzig, 5th edition, 1736) and in *Ausführliche Nachricht*, §23.

[10] In this and in all subsequent quotes, emphasizing italics are also in the original. Since Wolff uses italics very liberally, I occasionally omit them in the translation. Unless I refer to a translation into English, the translations are my own.

[11] Wolff, *Philosophia rationalis sive logica*, §519.

At any rate, Wolff does not seem too concerned about what the bearers of modal properties are. Soon he applies the notions of possibility and impossibility to definitions and hypotheses (§81) and to geometrical propositions. When he comes to the topic of necessity, the paradigmatic instances he offers are propositions: that a rectilinear triangle has three angles which are equal to two right ones, that two times two are four, that a being from itself (*ens a se*) exists (§279).[12]

The definition of necessity holds little surprise: 'that whose opposite is impossible, or involves a contradiction [§79], is called *necessary*' (§279). This definition uses the same ingredients as that of impossibility, and additionally the term 'opposite'. Thorough as always, Wolff supplies a definition of 'opposite' at the start of his chapter on necessity (§272): 'By *opposites* we understand here those which involve each other's negation.' Although Wolff does not make that explicit, it follows from his acccount that the possible, the impossible and the necessary are related as in the traditional modal square of opposition.

Given the interdefinability of possibility and necessity, one might expect the corresponding sections to be next to each other. However, the *Ontology* separates them by about 136 pages, and four chapters on topics that are not obviously related to them—on the determinate and indeterminate, on the notion of being, on identity and similarity, and on the singular and universal. The two chapters on the modal notions even fall under different larger headings (sections, in this case): possibility is treated in 'On essence, existence, and related notions', necessity in 'On the general affectations of things'. Not surprisingly, this results in much repetition. It also seems to have made it easier for apparent discrepancies to go unnoticed.

Wolff defends his definitions by appeal to mathematical practice:

> I have derived this notion from the examples of the arithmeticians and geometers, and would not have adopted it except for the reason that I have observed it to be in conformity with those. For everyone unanimously agrees that the truths contained in numbers and figures are necessary. (§279)

When arguing by *reductio ad absurdum*, mathematicians prove a theorem by deriving a contradiction from its opposite, which is thereby shown to be impossible. This practice presupposes that the possible does not involve a contradiction, and thus one direction of the biconditional that is entailed by Wolff's definition. Wolff is less explicit about why he thinks that not involving a contradiction is sufficient for possibility.

He further recommends his definition as being in conformity with ordinary ways of talking (§326), and with the received view among philosophers (§327). To back up this last point, he mentions Aquinas' account of the necessary as that which could not be different, and claims that this is accepted by Protestant theologians too.[13] By conformity, he presumably means extensional agreement: the same classification into

[12] In yet other places (§304, for example), the bearers of necessity are taken to be properties.

[13] §293 suggests that he has Scherzer in mind.

necessary and non-necessary is effected. He does not think that the folk, or other philosophers, have given the same definition.

Definitions of modal notions in terms of contradiction are often attributed to Leibniz. However, Wolff does not mention Leibniz in this connection, despite repeatedly giving him credit for certain other doctrines, such as the principle of sufficient reason or the doctrine of preestablished harmony. But Wolff did not claim originality for his definition, and he may have thought that the account was not original with Leibniz either. In fact, accounts along these lines were given long before Leibniz.[14]

So much for the wording of Wolff's definitions. In the next section, I shall turn to a discussion of how the definitions are to be understood.

## 2.2 Involving a Contradiction: A Closer Look

What is it for a proposition to involve a contradiction? Unfortunately, Wolff's chapters on modal notions do not give us an explicit answer. The interpreter needs to infer Wolff's understanding of the definitions from the way he applies them, as well as turn to other chapters.

Wolff discusses contradictions in his chapter on the principle of non-contradiction. 'A contradiction is contained in two propositions such that the same thing is denied to be by one which is affirmed to be by the other' (§31). This suggests the following slightly more precise version of the definition of impossibility (with 'U' for 'Unmöglichkeit', the German word for impossibility):

U0 Proposition $p$ is impossible iff for some $q$, $p$ involves both $q$ and $\sim q$.

Even taking U0 for granted, we are still left with the interpretive question what it is for a proposition to involve another proposition.

From the standpoint of contemporary philosophy, we can distinguish various ways in which we might understand involvement. One is in terms of counterfactuals: $p$ involves $q$ iff had $p$ been the case, then $q$ would have been the case. Combined with U0, this leads to an account of modality of the kind suggested in Lewis and developed in Williamson.[15] However, I am not aware of any evidence that this is what Wolff had in mind.

On another interpretation, $p$ involves $q$ iff $q$ is contained in an analysis of $p$. This seems quite plausible: Wolff uses the German word for 'to contain' (*enthalten*) as a synonym for the Latin *involvere*,[16] and it is natural to think of containment as a

[14] In her *Leibniz' Doctrine of Necessary Truth* (Garland, New York and London: Harvard Dissertations in Philosophy, 1990), M. D. Wilson identifies traces of this view in Aristotle, Aquinas, Nicholas of Autrecourt, and Hobbes.

[15] Lewis, *Counterfactuals* (Oxford: Blackwell Publishers, 1973), p. 22; Williamson, *The Philosophy of Philosophy* (Oxford: Blackwell, 2007).

[16] Wolff, *Der Vernüntigen Gedancken von Gott, der Welt und der Seele des Menschen, auch allen Dingen überhaupt, anderer Theil, bestehend aus ausführlichen Anmerckungen*, §6.

matter of conceptual analysis.[17] Arguably, Kant understood Wolff along these lines in his influential criticism of the latter's account of possibility. Kant claimed that a figure enclosed within two straight lines (*bilineum rectilineum*)—one of Wolff's favourite examples of an impossibility—does not involve a contradiction:

> It is, indeed, a necessary logical condition that a concept of the possible must not contain any contradiction; but this is not by any means sufficient to determine the objective reality of the concept, that is, the possibility of such an object as is thought through the concept. Thus there is no contradiction in the concept of a figure which is enclosed within two straight lines, since the concepts of two straight lines and of their coming together contain no negation of a figure.[18]

On Kant's view, intuition is needed to establish that there is no such thing, not merely an analysis of concepts.

While presumably correct, this interpretation is not too helpful for the contemporary reader as it stands. The notion of analytic containment is itself too contested today. Among those who deem it legitimate, most will side with Kant and deny that it applies in the above case. The interpretation would need to be supplemented by an account of how Wolff thought of analytic containment—something that cannot be attempted here.

On a third possible interpretation, involvement is cashed out in terms of apriority: $p$ involves $q$ iff 'if $p$ then $q$' is a priori. This account is reminiscent of David Chalmers' discussion of negative conceivability, which he takes to be cointensional with possibility.[19] According to Chalmers, a sentence or thought token $p$ is negatively conceivable iff $\sim p$ is not a priori. He takes the notion of apriority as primitive, but elucidates it via the notion of ideal reflection. This third interpretation fits well with Wolff's pervasive use of geometrical examples. Furthermore, since Wolff does not always distinguish sharply between the modal and the epistemic, the notion of apriority need not be out of place in an account of necessity. For now, I shall work with the hypothesis that 'involvement' can be understood in this way. I shall return to the interpretation of involvement in section 2.5.

So far, I have taken U0 for granted, and considered different interpretations of 'involves'. As it turns out, many passages cannot be accommodated by any plausible account of 'involves'. Perhaps U0 should be given up, then. There is indeed evidence that Wolff used 'involves a contradiction' to mean what we might express by 'involves a falsehood'. The definition of impossibility would then boil down to the following:

U1 Proposition $p$ is impossible iff $p$ involves some false proposition.[20]

[17] I am grateful to the anonymous referee for pointing this out.

[18] Kant, *Critique of Pure Reason*, trans. N. Kemp Smith (London: Macmillan, 1959), B268.

[19] Chalmers, 'Does Conceivability Entail Possibility?' in *Conceivability and Possibility*, edited by T. S. Gendler and J. Hawthorne (Oxford: Oxford University Press, 2002).

[20] Alternatively, one could retain U0 and claim that 'involvement' is disjunctive: $p$ involves $q$ iff $p$ involves $q$ in the ordinary sense—whatever that is—or $q$ is true. There are different ways to slice the pie.

One passage that fits U1 better than U0 occurs in Wolff's illustration of his definition of impossibility:

> A figure enclosed within two straight lines involves a contradiction, for example, because it supposes that two straight lines can be contained between the same points, which contradicts the true proposition that nothing but a unique straight line is contained between two points. Hence a figure enclosed within two straight lines is impossible. (§79)

Given U0, one would expect that Wolff might try to derive two different propositions, one of which is the contradictory of the other. Instead, he merely derives one and notes that it contradicts a true proposition.

In another passage, two paragraphs after the definition of impossibility, Wolff is even more explicit:

> Since that involves a contradiction, from which one can derive that it either contradicts itself or any true proposition; that is impossible, from which can be concluded that it either contradicts itself or any true proposition. (§81)

Further support for U1 over U0 is found in Wolff's discussion of the principle of non-contradiction, which precedes the chapter on possibility. He claims that when some *A* is *B* and some *A* is not *B*, then 'Every *A* is *B*' involves a contradiction (§38). As an example, he offers the sentence 'Every planet stands in opposition to the sun', which is falsified by Mercury and Venus.

The interpretation U1 also has the advantage of allowing a charitable interpretation of some of Wolff's putative proofs. In this context, I would like to look at two of his statements of particular interest for the student of the history of modal thinking: his theses about the modal status of modal claims themselves. Before the inception of modal logic early in the twentieth century, such theses did not receive much discussion. While Wolff was not the first to assert relevant principles, he is likely to have arrived at them independently.[21] He states theses which look strikingly similar in structure to principles 4 and 5 of modern modal logic (the characteristic axioms of the systems S4 and S5):

4' What is impossible is necessarily impossible. (§287)

5' What is possible is necessarily possible. (§286)

Despite the similarity, the attribution of axioms 4 and 5 to Wolff requires qualification. In modern modal logic, the significance of principles that might be expressed by the above words depends partly on logical background assumptions. A schema of the form $\sim\Diamond p \rightarrow \Box\sim\Diamond p$—symbolic for 'if *p* is impossible, it is necessarily impossible'—can be instantiated by any *p*, even one that contains modal vocabulary itself. It is not clear whether this is Wolff's intention. We need to recall that he had treated

[21] See I. Boh, 'Epistemic and Alethic Iteration in Later Medieval Logic', *Philosophia Naturalis* 21 (1984): 492–506, for references to earlier discussions, which were not widely available then, and still are not today. *Pace* some commentators, I do not think that voluntarists about eternal truths, such as Descartes, made a claim about repeated application of the same kind of modality.

possibility and impossibility in an earlier chapter. In the chapter on necessity, he states that whichever one of these properties a proposition had, it had it necessarily. He does not explicitly apply either possibility or necessity to itself: there is no principle to the effect that the necessary is necessarily necessary, or the possible impossibly impossible.[22]

Assuming U0, we might expect an argument for 4' to show that whenever *p* involves a contradiction, so does '*p* does not involve a contradiction'; and an argument for 5' to show that whenever *p* does not involve a contradiction, '*p* involves a contradiction' does.[23] However, Wolff does not proceed in this way. He offers the following putative proof of 4':

> For what is impossible involves a contradiction. Therefore, since it cannot in any way happen that the same both is and isn't, it cannot happen either that the impossible does not involve a contradiction. For the opposite of the impossible is the possible, and hence the impossible is necessarily such. (§287)

It is certainly tempting to accuse Wolff of having a muddled understanding of what it takes to apply one modal concept to another one, or of committing a version of the fallacy of confusing the necessity of the consequence with the necessity of the consequent—roughly, to infer 'if the impossible involves a contradiction, then it is necessary that the impossible involves a contradiction' from the necessity of the tautological conditional 'if the impossible involves a contradiction, then the impossible involves a contradiction'. But the interpretation of 'involves a contradiction' suggested can rescue him from that charge. It allows for the following reconstruction of the argument. Suppose that *p* is impossible. Then it is true that *p* is impossible. Consider now the opposite of *p*'s being impossible, namely its being possible. Its being possible involves itself—as surely everything does—and thus involves a false proposition. By U1, the opposite of *p*'s being impossible is impossible. Therefore, it is necessary that *p* is impossible.

Wolff's putative proof of 5' is analogous:[24]

> for what is possible does not involve a contradiction. Therefore, since it is impossible that the same both is and is not, it cannot happen that the possible involves a contradiction. The opposite of the possible is indeed the impossible, and consequently that which is possible is necessarily possible. (§287)

To reconstruct this argument, suppose that *p* is possible. It is then true that *p* does not involve a contradiction. Thus the supposition that *p* involves a contradiction involves

[22] Though J. N. Findlay (*Kant and the Transcendental Object: A Hermeneutic Study* (Oxford: Oxford University Press, 1981)) thinks that this is implied: Wolff 'implies, but does not clearly say, that the necessary is also the necessarily necessary' (p. 41).

[23] For a relevant contemporary discussion of such iterations, see J. P. Burgess, 'Which Modal Logic is the Right One?', *Notre Dame Journal of Formal Logic* 40 (1999): 81–93.

[24] §38 of Wolff's *Vernünftige Gedancken von Gott, der Welt und der Seele des Menschen, auch allen Dingen überhaupt* (Halle, 1751; reprinted Hildesheim: Georg Olms, 1983), offers a similar argument.

a falsehood, namely itself. By U1, it is impossible that *p* involves a contradiction. Hence its opposite, that *p* does not involve a contradiction, is necessary. Hence it is necessary that *p* is possible.

While it avoids attributing a gross fallacy to Wolff, the interpretation does come with a significant cost: it seems to commit Wolff to fatalism, the view that all truths are necessary. For if *p* is true, then ~*p*, being false and involving itself, involves a falsehood. By U1, ~*p* is impossible; hence its opposite, *p*, is necessary. An analogous principle will hold for possibility: if *p* is possible, then it does not involve a false proposition; hence *p* is not false, and, given the law of excluded middle, it is true.

As it turns out, there is some evidence that Wolff embraced this consequence. In the paragraph following those asserting the necessity of possibility and of impossibility, respectively, he claims:

Anything is necessarily while it is. (§288)

The proof for this takes a similar pattern as those we have just seen.[25] However, it should be noted that the principle is not an unambiguous affirmation of fatalism, because of the temporal qualification. Everything is said to be necessary while it exists (*dum est*), leaving it open that it was not or will not be necessary.

There is considerable evidence that Wolff rejected fatalism. In many passages, he asserts that the possible need not be actual. I shall restrict myself to a representative sample here. In the relatively early *German Logic*, after clarifying that 'everything that can be I call possible, whether it is actual or not' (§3), he claims:

... it is thus certainly more than clear, that something is not yet because it is possible, and that one cannot infer from the possibility alone that something will be. For if I recognize that something is possible, I cannot assume, for that reason, that it is in reality, or has been before, or will come henceforth.[26]

Similarly, in 1740:

By explicating the possible in terms of that which does not contain a contradiction in itself, I take the word in a wide sense, such that many things that never become real fall under it. And in this respect I distance myself from Spinoza and other fatalists, who only take that to be possible which also is real at some time.[27]

[25] At least one sympathetic commentator on Wolff's *Ontology*, Hans Pichler, notices both Wolff's commitment to fatalism and his apparent acceptance of it in §288: 'Since everything which contradicts a true sentence is impossible, everything, as it is, must be necessary, for the opposite claim would contradict the fact, and therefore involve a contradiction.... Surprisingly, Wolff draws this consequence himself.... This concept of necessity is entirely devoid of significance.' (Hans Pichler, *Über Christian Wolffs Ontologie*. Leipzig: Verlag der Dürr'schen Buchhandlung, 1910, p. 38). Pichler suggests discounting Wolff's unqualified modal notions in favour of the qualified ones, which are discussed in section 3, below.

[26] Wolff, *Vernünftige Gedanken von den Kräften des menschlichen Verstandes* (Halle, 1754; reprinted Hildesheim: Georg Olms, 1965), §13.

[27] Wolff, *Der Vernünftigen Gedancken von Gott, der Welt und der Seele des Menschen, auch allen Dingen überhaupt, anderer Theil, bestehend aus ausführlichen Anmerckungen*, §6.

Finally, in the *Cosmology*, written after the *Ontology*, he explicitly asserts that there are other possible worlds, or universes.[28]

To summarize this section: there seems to be a serious tension in Wolff's account of modality. On the one hand, he seems to hold that everything that contradicts a truth is impossible. This would commit him to the claim that every falsity is impossible. On the other hand, he asserts that the possible need not exist.[29]

## 2.3 Absolute and Hypothetical Necessity

So far, I have examined Wolff's basic modal notions, and the interpretive difficulties they raise. It might be thought that these difficulties can be solved by appealing to certain distinctions between kinds of necessity and possibility that Wolff draws. Indeed, it seems that he himself took such distinctions to be the key to avoiding fatalism: 'A great deal hangs on one's properly conceiving of this distinction [between that which is absolutely (*schlechterdings*) possible and that which is possible in this world], in so far as one wishes to get by in the difficult matter of necessity and contingency, and to thoroughly respond to the fatalists'.[30] Even an apparently contingent fact may be necessary, in some sense. What matters is that there is also a pertinent sense of necessity in which it is not necessary.

Wolff characterizes the kinds of necessity and possibility in various different ways, and also uses different labels. I shall focus here on two pairs: absolute and hypothetical necessity as defined in the *Ontology*, and intrinsic and extrinsic possibility as defined in the *Cosmology*. In this section, I discuss how Wolff characterizes these notions. In the next one, I shall ask whether they can do the philosophical work he intends them to do.

In §302, Wolff defines absolute and hypothetical necessity. The marginal title of §302 is 'difference between absolute and hypothetical necessity', suggesting a definition into genus and specific difference. So hypothetical necessity is indeed a kind of necessity. The previous paragraphs (§§279–299) had been concerned with a generic notion of necessity. This becomes clear when Wolff goes on to discuss, of the previously asserted necessities, 'what has absolute necessity, and whose necessity is, in contrast, merely hypothetical' (§ 302). For example, after trying to prove in §299 that the essences of species are necessary, he argues for their absolute necessity in §303.

[28] Wolff, *Cosmologia Generalis* (Frankfurt and Leipzig, 1737; reprinted Hildesheim: Georg Olms, 1964), §§48, 100, 101.

[29] One could try to interpret Wolff as tacitly restricting his propositional quantifier: something is impossible if there is a truth *of a certain sort* that it contradicts. The restriction might even be contextual: the relevant truths could be taken to be those that Wolff has previously established. Such an interpretation fits some of Wolff's texts. However, it has trouble accommodating the proofs of the necessity of possibility and impossibility, quoted above. Not all true possibility and impossibility claims have been established in Wolff's works, after all.

[30] Wolff, *Der Vernünftigen Gedancken von Gott, der Welt und der Seele des Menschen, auch allen Dingen überhaupt, anderer Theil, bestehend aus ausführlichen Anmerckungen*, §6.

As we have seen, the necessary is that whose opposite involves a contradiction. Absolute and hypothetical necessity differ as to the source of the contradiction:

> That which is in itself or absolutely considered such that its opposite is impossible, or involves a contradiction, is called *absolutely necessary*. But that whose opposite is impossible, or involves a contradiction, only under a hypothesis, or under a certain given condition, is *hypothetically necessary*. (§302)

In the last section, I argued that for Wolff, something is necessary if its opposite involves a contradiction with any truth. Presumably, the distinction between these kinds of necessity turns on whether such truths need to be themselves involved in the opposite of the putatively necessary, or whether they could be truths about something external.

Wolff's definition makes the two kinds exclusive: nothing is both absolutely and hypothetically necessary. This is one respect in which Wolff's conception of absolute necessity is different from the contemporary one, on which a proposition that is absolutely necessary enjoys any other kind of necessity as well.

In his previous paragraph, Wolff has defined the difference between something being considered absolutely or under a given hypothesis. We consider a thing absolutely if we only pay attention to its essence, or definition. Wolff explains that this kind of necessity is also called 'geometrical' or 'metaphysical' (§304). I shall discuss this notion of absolute necessity below.

The notion of hypothetical necessity appears to be more obscure. We consider something under a hypothesis when 'beyond the essence we presuppose other determinations at the same time, which are not yet posited when the thing is posited, but whose being posited is at least not incompatible with it' (§301). This definition is not too helpful, since it does not constrain what these other determinations are. For all we have been told, even falsehoods could be hypothetically necessary. We later learn that modi are hypothetically necessary (§306). In the *Cosmology* (§102), Wolff asserts that what happens in the world is hypothetically necessary. Perhaps we can understand the hypothetically necessary to be that which is necessary but not absolutely so, as suggested by §319 of the *Ontology*.

The distinction between absolute and hypothetical necessity has no analogue in the chapter on possibility of the *Ontology*. However, distinctions among kinds of possibility can be found in other works, such as the earlier *Notes on the German Metaphysics*[31] and the later *Cosmology*. In the latter, Wolff defines a notion of intrinsic possibility:

> *Intrinsically possible* is that which considered in itself is in a certain way, that is, which considered in itself does not involve a contradiction.[32]

[31] Wolff, *Der Vernünftigen Gedancken von Gott, der Welt und der Seele des Menschen, auch allen Dingen überhaupt, anderer Theil, bestehend aus ausführlichen Anmerckungen*, §6.

[32] Wolff, *Cosmologia Generalis*, §111.

It appears that absolute necessity and intrinsic possibility are duals of each other.

Something is intrinsically possible, roughly, if it is not ruled out by its own nature, regardless of the environment. Wolff contrasts intrinsic possibility with extrinsic possibility. We might expect something to be extrinsically possible just in case it is not ruled out by the nature of the environment. However, this is not how Wolff understands that notion:

> *Extrinsically possible* is that which has a determinate cause in the visible [adspectabilis] world, that is, what is able to exist in it. *Extrinsically possible things* are also called *possibilia of this world.*[33]

Given that causes were widely taken to be necessitating at that time, the notion defined here would arguably be more aptly called 'extrinsic necessity'. Wolff confirms the impression that it functions like a necessity-operator by claiming that whatever is extrinsically possible is hypothetically necessary (§114). By way of further clarification, he adds that only what has existed, or is present, or will at some point exist, is extrinsically possible in this world (§112).

The duality of intrinsic possibility and absolute necessity may suggest that extrinsic possibility is the dual of hypothetical necessity—which I stipulatively call 'hypothetical possibility'. However, even a brief look at the definitions shows that they appeal to rather different sets of notions. While extrinsic possibility is characterized in metaphysical terms—causation, determination, the world—hypothetical possibility seems to be a broadly logical notion. Closer inspection reveals that they need not be coextensional either. Suppose that $p$ is absolutely necessary. Then it is not hypothetically necessary. Hence its negation $\sim p$ is hypothetically possible. But presumably, $\sim p$ does not have a cause in the visible world, and is thus not extrinsically possible. So the notions have different extensions. Can we at least conclude that if $p$ is hypothetically possible but not absolutely necessary, it is extrinsically possible? We can, provided that the hypotheses in question include all truths, and that a version of the principle of sufficient reason guarantees that everything which is not absolutely necessary has a determinate cause in the visible world.

## 2.4 Fatalism versus Contingentism

The distinction between absolute and hypothetical necessity is supposed to help in avoiding fatalism. The idea is clear: even if all truths are necessary, some are not absolutely necessary; and fatalism does not follow unless all truths are absolutely necessary.

On a closer look, the relevance of the distinction for the question of fatalism is not so clear. To explain why, I shall consider Wolff's account of a modal notion which I have not yet discussed, namely contingency.

[33] Wolff, *Cosmologia Generalis*, §111.

Wolff defines something as contingent if it is not necessary, or if its opposite does not involve a contradiction (§294). He offers the following as paradigmatic examples: the heat of the stone, the erudition of the man, and human actions. These are not examples of concepts, but of modes of things—perhaps corresponding to tropes in today's metaphysical terminology. Presumably, those modes are contingent just in case the singular propositions ascribing them to their bearers are.[34]

Wolff's opposition to fatalism is usefully stated in terms of contingency. He may be willing to accept that everything enjoys a necessity of some kind. But he would not wish to deny any contingency in the world. In one place, he characterizes his dispute with Spinoza about modality as between contingentists (*contingentiarios*) and fatalists.[35]

But suppose that U1 (section 2.2) is the correct interpretation, such that for Wolff, the—generically—necessary is that whose opposite involves a contradiction with some truth. Then it seems that every truth is necessary, and that the definition of the contingent as the non-necessary entails that there are no contingent truths. The distinction between absolute and hypothetical necessity—which appears after the definition of contingency in the *Ontology*—has no bearing on this.

This is not the conclusion that Wolff draws, however. He claims that 'only absolute necessity is opposed to contingency..., or removes contingency; hypothetical [necessity] does not' (§319). When arguing for this, he implicitly takes the contingent to be that which is not absolutely necessary, rather than that which is not necessary, as he had originally defined it. Indeed, he occasionally slips into talk about something's being 'contingent in itself' (§318).[36]

Suppose that we disregard the untidy introduction of the notion, and understand contingent truths to be those that are not absolutely necessary. Does Wolff manage to distance himself from fatalism?

In one respect, he does. He can assert that some things are possible—intrinsically possible—but not actual, and that some truths are contingent. So he seems to deny the letter of fatalism. But let us recall what Friedrich Wilhelm was concerned about: that if one of the Potsdam Giants deserts, he cannot be punished because his desertion was necessary. Wolff can respond that the desertion is not absolutely necessary. It is a mode, not a consequence of the essence of the tall lad that he deserts. The desertion is, however, hypothetically necessary. If that is a genuine kind of necessity, it is one that has its source outside the essence of the thing concerned.

[34] Today, contingency is taken to entail possibility, not just non-necessity, unlike in Wolff's account. But I shall not dwell on that here.

[35] Wolff, *De differentia nexus rerum sapientis et fatalis necessitatis* (Halle, 2nd edition, 1737), §8.

[36] Wolff could get the same result by retaining the definition of the contingent as the non-necessary, but taking necessity to be what he calls 'absolute necessity' in the *Ontology*. He seems to have been aware of the advantages of such a terminological choice. In §17 of his *Der Vernünftige Gedancken von Gott, der Welt und der Seele des Menschen, auch allen Dingen überhaupt, anderer Theil, bestehend aus ausführlichen Anmerckungen*, Wolff writes that apart from the absolutely necessary, what is called necessary 'does not deserve that name', and that it would be better if one had refrained from naming it so.

This can easily be taken to mean that the grenadier was compelled by outside forces to desert. The thought that the desertion was due to external rather than internal factors surely makes the argument against punishment stronger, rather than weaker. Self-determination is often taken to be an important condition for freedom or some of the moral features that are related to it, such as blameworthiness and responsibility.[37] Sometimes self-determination is even taken to be a sufficient condition, even in the absence of alternative possibilities of the self-determining agent. Perhaps the king would have found it more reassuring to be told that the desertion was absolutely necessary.

The topic of freedom and responsibility is a notorious minefield. It would be risky to rest a case against an interpretation on considerations such as the above. However, Wolff's account is independently problematic. For it is not clear that what he calls 'absolute necessity' is really a kind of necessity; and if it is not, he has not articulated a sense of necessity in which a truth may fail to be necessary.

One problem with the notion of absolute necessity concerns its proper elucidation. Given a certain proposition, what is it that we are supposed to consider in itself? Whose essences are we to consider? In a singular proposition, such as 'the sun moves' or 'this stone is hot', we presumably consider the essence of the subject term and ask whether it contradicts the predicate term. But it is not clear how this account is to be generalized. One option would be to take propositions to have constituents, and to consider the essences of all the objects that are constituents of the proposition.[38]

But however exactly it is spelled out, it seems clear that absolute necessity will lack certain crucial inferential features that we expect a necessity operator to have. It is a central principle of modal logic that if both $p$ and 'if $p$ then $q$' are necessary, so is $q$.[39] But this principle will fail for absolute necessity. This has been observed by Robert M. Adams in his excellent discussion of Leibniz's analogous notions of a thing possible in its own nature and of necessity through itself.[40] For example, it may be absolutely necessary that God exists, and likewise that if God exists, this world is actual—because it follows from its essence that it is the best, and because it follows from God's essence that He actualizes the best. However, it is arguably not absolutely necessary that this world is actual, since its being actualized is not part of its essence.

In summary, then, it is not clear whether the distinction between absolute and hypothetical necessity helps Wolff articulate a contingentist position.

[37] As J. Wisdom writes: 'Unless our behaviour is partly determined by our nature, we are never to blame'; 'Freedom, Causation, and Preexistence: an Excerpt from *Problems of Mind and Matter*' in *Metaphysics: the Big Questions*, edited by P. van Inwagen and D. W. Zimmerman (Malden: Blackwell, 2008), 432–41, p. 433.

[38] See K. Fine, 'The Logic of Essence', *Journal of Philosophical Logic* 24 (1995): 241–73 for such an approach.

[39] This axiom is standardly called K.

[40] Adams, *Leibniz: Determinist, Theist, Idealist* (Oxford: Oxford University Press, 1994), pp. 12–20. See also the discussion of the logic of local possibility in F. Correia '(Finean) Essence and (Priorean) Modality', *dialectica* 61 (2007): 63–84.

## 2.5 Wolff's Analysis of Propositions and the Possibility of Aliens

In section 2.2, I argued that for Wolff, 'involving a contradiction' does not mean what we would normally take it to mean, and suggested the interpretation U1, which leads to fatalism. However, things are more complicated, as will emerge in this section.

As noted already, Wolff discusses contradictions extensively in his elucidation of the principle of non-contradiction. Earlier, I used a quote from that chapter in support of U1 over U0. It turns out that Wolff says other things there that fit neither U0 nor U1. Moreover, he sketches an analysis of propositions which casts doubt on our working hypothesis that 'involves' is to be understood, roughly, as a priori entailment. This allows for an interpretation on which Wolff avoids fatalism.

Before presenting this interpretation, I wish to discuss a methodological question. The interpretation rests heavily on what Wolff writes in a chapter on the principle of non-contradiction, which precedes any explicit discussion of modal terms. Wolff is not generally averse to recycling his material in slight variations, and has a habit of referencing earlier paragraphs copiously. Strikingly though, later parts of the *Ontology* do not restate his analysis of propositions, and what it takes for propositions to contradict each other, or refer the reader back to the relevant paragraphs in the chapter on the principle of non-contradiction. This raises the question whether it is legitimate to read the chapters on modality in the light of the earlier one.

There are two reasons why this seems defensible to me. The first is that Wolff explicitly claims that the discussion of contradiction matters for his subsequent account of impossibility:

> It is useful for us ... to make the concepts of a contradiction clearer, since it is an ingredient of other ontological concepts. As it will soon be established, an impossibility cannot be argued for unless one can state a contradiction. (§32)

The second is that Wolff took his analysis of propositions to be a great achievement, and praised its usefulness:

> Leibniz ... admonished that there is more need for light in first philosophy than in mathematics, and held that a certain unique method for putting forward theses is necessary, with whose help questions are to be resolved by the euclidean method, in the manner of a calculus.... But that unique method to propose theses is the knot that none of the philosophers has so far solved, and not even Leibniz has indicated, still less taught, how it is to be solved. However, we do not doubt at all that with the benefit of the analysis of universal and particular propositions into singular ones, in the theory of contradiction, we have solved this knot in the most simple manner. (§51)

This suggests that this analysis was intended to inform further theorizing in first philosophy.

So what is Wolff's theory of contradiction? One of his striking claims is nicely captured by the marginal title of §33: 'That a contradiction genuinely obtains only between singular propositions'. We can combine this with the view that the impossible is that which involves any falsehood to obtain the following interpretation:

U2 Proposition *p* is impossible iff *p* involves some false singular proposition.

To motivate the claim that a contradiction is always between singular propositions, I shall introduce Wolff's analysis of propositions. For this, in turn, some background from traditional logic is needed.

Following the tradition, Wolff divides propositions according to their quantity into universal, particular, and singular. Paradigmatic instances of each group are 'all planets are dark', 'some planets are not dark', and 'Jupiter is dark', respectively. Still following tradition, he also divides them into affirmative ones—'Jupiter is dark'—and negative ones—'Jupiter is not dark'.

According to the received view in Wolff's day, as captured by the square of opposition, universal affirmatives contradict particular negative propositions, and likewise, universal negatives contradict particular affirmative ones. Typically, logic texts offered no explanation of why these relationships hold. Wolff, however, thought he could provide an explanation:

> In the case of a contradiction between a universal affirmative and a particular negative, and between a universal negative and a particular affirmative proposition, the contradiction obtains really between singular propositions with the same subject and predicate. Therefore, a contradiction obtains always between singular propositions. (§33)

Wolff's idea is that universal and particular propositions are to be analysed as complexes of singular ones. About universal ones, he writes:

> He who claims that all planets are dark certainly claims that Saturn is dark, that Jupiter is dark, that Mars is dark, and so forth. And so a universal proposition is a complex of several singular propositions, and mostly there is no way to enumerate them. (§33)

This suggests the following account: the universal proposition 'Every *A* is *B*' is the complex consisting of all singular propositions predicating *B* of some thing which is actually an *A*. It is tempting, of course, to read 'complex' as 'conjunction', but this would be anachronistic.

Even though Wolff does not say so, it is natural to hold that 'Every *A* is *B*' involves all the singular propositions which are its constituents. This assumption forces us to modify our working assumptions that involvement is a priori entailment. Since it is not a priori that Jupiter exists, and is a planet, 'All planets are dark' does not a priori entail 'Jupiter is dark'. Nonetheless, the former involves the latter, on the current interpretation.

Given such a bridge principle between analysis and involvement, we can derive consequences for the modal status of certain propositions from this theory. Suppose that the universal proposition 'Every *A* is *B*' is false. Then some *A*—say $a_1$—is not *B*. So '$a_1$ is not *B*' is a true singular proposition. But 'Every *A* is *B*' involves '$a_1$ is *B*', and is impossible by U2. In this argument, nothing depends on the proposition being affirmative. So any false universal proposition is impossible. This conforms with what we have already seen in section 2.2 in connection with the example 'Every planet stands in opposition to the sun.'

Next, suppose that the particular proposition 'Some *A* is *B*' is true. Then its opposite, the universal proposition 'No *A* is *B*', is false. By the argument from the last paragraph, it is impossible. So its opposite, 'Some *A* is *B*' is necessary. In general, any true particular proposition is necessary.

So far, it does not seem that the theory opens the door to contingency. But things get interesting once we consider false particular propositions. Suppose 'Some *A* is *B*' is false. Let $a_1, \ldots, a_n$ be all the *A*s. Then '$a_1$ is *B*', '$a_2$ is *B*', ..., '$a_n$ is *B*' are false singular propositions. Does 'Some *A* is *B*' involve any of them? The natural answer is no. For easy reference, I label that thesis 'NIT' ('non-involvement thesis').

NIT False particular propositions do not involve any false singular ones.

At any rate, 'Some *A* is *B*' does not entail '$a_1$ is *B*', say, even in conjunction with '$a_1$ is *A*'. So 'Some *A* is *B*' is false but not impossible, given NIT. Therefore, its opposite 'No *A* is *B*' is contingent if true. Again, it does not matter whether the propositions are affirmative or negative: 'Every *A* is *B*' is contingent if true, and 'Some *A* is not *B*' is still possible when it is false.

When discussing the consequences of this theory, I have presupposed that NIT should be attributed to Wolff. Is that attribution justified? Unfortunately, Wolff rather muddies the water when discussing how his analysis applies to particular propositions. Rather than assembling and commenting on various passages that appear to be inconsistent with each other, I shall only present a passage which supports the attribution:

> [T]he true particular proposition consists of mere true singular ones.... The following proposition serves as an example: *Some planets never stand in opposition to the sun*—namely the lower ones—which consists of two true singular propositions, to which it is equivalent: *Venus never stands in opposition to the sun*, and *Mercury never stands in opposition to the sun.* (§37)

The passage is only explicitly concerned with true particulars. It suggests that 'Some *A* is *B*', if true, is a complex consisting of all true singular propositions predicating *B* of an *A*. But it would seem to get things back to front if the analysis of a proposition were to depend on whether it is true. Presumably, the analysis should tell us under what conditions the proposition is true.

It is thus reasonable to assume that 'Some *A* is *B*' consists of all true singular propositions predicating *B* of an *A* even if it is false. But there are no such if 'Some *A* is *B*' is indeed false. So a false particular proposition gets analysed into a 'complex' of no singular propositions. In some respects, such an 'analysis' is surely problematic. But it does justify NIT: false particular propositions do not involve any false singular propositions, since they do not involve any singular propositions at all.

On this account, not all truths are necessary, and not all falsehoods impossible. So it seems that Wolff's theory of propositions helps him avoid fatalism, the claim that no proposition is contingent. It does so without reliance on the distinction between absolute and hypothetical necessity. Moreover, the account has the potential to vindicate Wolff's claim that not all possibles are actual. Presumably, a golden mountain is possible iff the proposition 'Some mountain is golden' is possible. As a false particular, that proposition is possible, as I have just explained.

The theory does face a problem, however: it over-generates possibilities. Some particular proposition, such as 'Some piece of wood is made of iron', or 'Some planet is not a planet', are impossible. But by NIT, they do not involve any false singular propositions, and are thus possible by U2. So the theory as reconstructed here cannot be right, and would at best need amendment.

Supposing that this issue could be solved—I shall not discuss how this could be done—the resulting view could be modelled using possible worlds, as follows. Every possible world contains all the actual individuals, with all the properties that they actually have. In some possible worlds, there are additional non-actual individuals, so-called aliens, some of which have properties that are not actually instantiated. The fact that the view can be modelled in a possible-worlds setting shows that it satisfies the standard principles of modal logic.[41] So it gives us concepts of possibility and necessity that avoid fatalism and are not logically revisionary, unlike the concepts of intrinsic possibility and absolute necessity.

Whatever the technical merits and flaws of the account, it would not have been likely to satisfy King Friedrich Wilhelm. If a given soldier *a* deserts, then the singular proposition '*a* deserts' is true. The account then entails that '*a* deserts' is necessary, which the king took to have unacceptable consequences concerning freedom and punishment. While on my interpretation, Wolff's theory is not fatalistic in letter—it allows for some propositions to be contingent—it is arguably fatalistic in spirit, since true singular propositions about people's actions turn out to be necessary.

## Bibliography

Adams, R. M., *Leibniz: Determinist, Theist, Idealist* (Oxford: Oxford University Press, 1994).

Boh, I., 'Epistemic and Alethic Iteration in Later Medieval Logic', *Philosophia Naturalis* 21 (1984): 492–506.

Burgess, J. P., 'Which Modal Logic is the Right One?', *Notre Dame Journal of Formal Logic* 40 (1999): 81–93.

Chalmers, D. J., 'Does Conceivability Entail Possibility?' in *Conceivability and Possibility*, edited by T. S. Gendler and J. Hawthorne (Oxford: Oxford University Press, 2002).

Correia, F., '(Finean) Essence and (Priorean) Modality', *dialectica* 61 (2007): 63–84.

[41] The natural modelling of accessibility would be as follows: *w* accesses *w'* iff every singular proposition true at *w* is also true at *w'*. This would vindicate principle 4' discussed in section 2.2, but not 5'.

Findlay, J. N., *Kant and the Transcendental Object. A Hermeneutic Study* (Oxford: Oxford University Press, 1981).

Fine, K., 'The Logic of Essence', *Journal of Philosophical Logic* 24 (1995): 241–73.

Hinrichs, C., *Preussentum und Pietismus* (Göttingen: Vandenhoeck & Ruprecht, 1971).

Kant, I., *Critique of Pure Reason*, trans. N. Kemp Smith (London: Macmillan, 1959).

Lewis, D., *Counterfactuals* (Oxford: Blackwell Publishers, 1973).

Pape, I., *Tradition und Transformation der Modalität* (Hamburg: Meinerk, 1966).

Pichler, H., *Über Christian Wolffs Ontologie* (Leipzig: Verlag der Dürr'schen Buchhandlung, 1910).

Williamson, T., *The Philosophy of Philosophy* (Oxford: Blackwell, 2007).

Wilson, M. D., *Leibniz' Doctrine of Necessary Truth* (Garland, New York and London: Harvard Dissertations in Philosophy, 1990).

Wisdom, J., 'Freedom, Causation, and Preexistence: an Excerpt from *Problems of Mind and Matter*' in *Metaphysics: The Big Questions*, edited by P. van Inwagen and D. W. Zimmerman (Oxford: Blackwell, 2008), 432–41.

Wolff, C., *Ausführliche Nachricht von seinen eigenen Schriften, die er in deutscher Sprache heraus gegeben* (Frankfurt, 2nd edition, 1733; reprinted Hildesheim: Georg Olms 1973).

Wolff, C., *Philosophia rationalis sive logica. Praemittitur Discursus Praeliminaris De Philosophia in Genere* (Verona, 3rd edition, 1735).

Wolff, C., *Philosophia prima sive ontologia* (Frankfurt and Leipzig, 2nd edition, 1736a; reprinted Hildesheim: Georg Olms, 1962).

Wolff, C., *Vernünftige Gedancken von der Menschen Thun und Lassen* (Frankfurt und Leipzig, 5th edition, 1736).

Wolff, C., *Cosmologia Generalis* (Frankfurt and Leipzig, 1737; reprinted Hildesheim: Georg Olms, 1964).

Wolff, C., *De differentia nexus rerum sapientis et fatalis necessitatis* (Halle, 2nd edition, 1737).

Wolff, C., *Theologia naturalis*, vol. II (Frankfurt and Leipzig, 1737).

Wolff, C., *Der Vernünftige Gedancken von Gott, der Welt und der Seele des Menschen, auch allen Dingen überhaupt, anderer Theil, bestehend aus ausführlichen Anmerckungen* (Frankfurt, 4th edition, 1740; reprinted Hildesheim: Georg Olms, 1983).

Wolff, C., *Vernünftige Gedancken von Gott, der Welt und der Seele des Menschen, auch allen Dingen überhaupt* (Halle, 1751; reprinted Hildesheim: Georg Olms, 1983).

Wolff, C., *Vernünftige Gedanken von den Kräften des menschlichen Verstandes* (Halle, 1754; reprinted Hildesheim: Georg Olms, 1965).

# 3

# Modal Adventures between Leibniz and Kant

## Existence and (Temporal, Logical, Real) Possibilities

*Ohad Nachtomy*

'... *nisi* ... *Deus existeret, nihil possibile foret*' (GP VI 440)[1]

Kant's refutation of the ontological argument marks a moment in the history of philosophy in which the notion of existence becomes independent from that of essence. Kant's refutation is based on denying the premise, held by all upholders of the ontological argument, that existence is an attribute or a predicate.[2] As Vilkko and Hintikka remark, 'if we examine what Kant meant, we can see that his claim was far stronger than what the slogan "existence is not a predicate" expresses. He argued that existence cannot even be a part of the force of a predicate term.'[3] For all its novelty and consequences for theology[4], Kant's point seems rather straightforward.[5]

[1] GP VI 440: for the sense of this abbreviation and of others in this chapter, see the note on the method of citation at its end. In rough translation, 'unless God existed, nothing would be possible'. Leibniz's dictum is also echoed in his *Theodicy* §184: '*Sans Dieu, non seulement il n 'y auroit rien d'existant mais, il n 'y auroit rien de possible.*' 'Without God, not only would there be nothing existing but nothing would be possible either.'

[2] I am most grateful to Reed Winegar and an anonymous referee from Oxford University Press for very perceptive comments and suggestions. This research was supported by grant 469/13 from the Israel Science Foundation.

[3] R. Vilkko and J. Hintikka, 'Existence and Predication from Aristotle to Frege', *Philosophy and Phenomenological* Research, LXXIII/2 (2006): 359–77, p. 367.

[4] Kant's refutation of the ontological argument was perceived as most destructive, not only for traditional theology but also for the traditional world-view at large. Heine spoke of Kant as the *Weltzermalemender*, 'the great destroyer in the kingdom of thought'. See reference in Allen W. Wood, *Kant's Rational Theology* (Ithaca, 1978), pp. 16–17 and 97.

[5] It is perhaps for this reason that, despite its enormous consequences, Kant hardly feels the need to argue for it in the first *Critique*. I am not claiming that Kant's argument is flawless. It is worth noting that many philosophers remain unconvinced; see, for example, Graham Oppy's *Ontological Arguments and Belief in God* (Cambridge: Cambridge University Press, 1995).

Existence, he says, does not add anything to the essence or to the concept of a given thing (A599/B626). 'The actual,' he writes, 'contains nothing more than the merely possible. A hundred actual dollars do not contain the least bit more than a hundred possible ones' (A599/B627; CE 567). Existence does not add anything to the content of a concept; it only indicates that something is actual rather than merely possible.[6] In other words, in denying that existence adds something to the content of a concept (his negative thesis) Kant affirms (his positive thesis) that ascribing existence to x only points to x's modal status or position.

At the turn of the nineteenth century Frege and Russell assimilated this way of viewing existence into formal logic by formalizing the notion of existence not as a predicate but rather through the usage of an existential quantifier. This has become the canonical way of formalizing and (as many following Quine believe) of analysing the meaning of existence. Vilkko and Hintikka put this point very eloquently thus: 'after Kant, existence was left homeless. It found a new home in the algebra of logic.... The orphaned notion of existence has found a new home in the existential quantifier.'[7]

This transformation in the conception of existence from its status as a predicate or attribute (as is explicit in, say, Descartes' versions of the ontological proof) to its interpretation as a modal judgement (or a position) in Kant is indeed dramatic. Such a transformation could not have taken place without a significant change in the philosophical background, as I shall argue in this paper, a background that is fascinating and complex indeed.

Accounts of this background often omit that, more than a century before Kant, Leibniz has already articulated a similar position regarding existence.[8] For Leibniz, 'existence' does not (and cannot) add anything to a concept of an individual, which, according to him, is already complete as a candidate for actualization. In his correspondence with Arnauld, for example, Leibniz states that individuals are already found fully formed (*toute formée*) as possibilities i.e., as complete concepts in God's mind, so that nothing could be added to their concepts (DM 13, and the correspondence with Arnauld). Such complete concepts include all the would-be activities and properties of individuals—past, present, and future—as predicates in the concepts God considers in his understanding as candidates for actualization. What Leibniz's God considers for actualization (i.e., whether to create or not) are possible individuals, seen as complete concepts whose predicates fully specify their natures prior to their creation.[9] It seems quite clear that, in this framework, existence is

[6] CPR, A599/B627.

[7] For further discussion, see: Risto Vilkko and Jaakko Hintikka, 'Existence and Predication from Aristotle to Frege', p. 359.

[8] Cf. H. Ishiguro, *Leibniz's Philosophy of Logic and Language* (Cambridge: Cambridge University Press, 1990), p. 192.

[9] God did not choose to create an 'Adam vague' (LR 87), that is, an indefinite notion of Adam which entails only general characteristics (conceived *sub ratione generalitatis*). Rather, God chose to create a specified and well-defined notion of Adam. Leibniz writes that, 'the nature of an individual ['which he finds

not included in the concepts of individuals themselves, seen as *per se* possibilities (that is, independently of the compossibility relations, which make up for possible worlds).[10]

Furthermore, Leibniz even argues that the conception of existence as a predicate can be reduced *ad absurdum*: 'if existence were anything other than what is demanded by essence (*essentiae exigentia*), it would follow that it itself would have a certain essence, or would add something new to things, concerning which it might be asked, whether this essence exists, and why it and not another' (A 6.4 1443; GP VII 195. See also A 6.4 762–3). If existence were to be one of the essential predicates of the complete concept of individuals, God's choice to create the best possible world would have been redundant, as such an individual would have to exist regardless of God's choice. Furthermore, in case 'existence' were one of the individuals' essential predicates, creation would seem to be a direct logical consequence of these concepts. Likewise, Leibniz's central thesis that the actual world is *contingent* upon God's choice to create *it*, rather than another possible world, would be vacuous, for creation would involve no choice or rational deliberation.

Indeed, with respect to the concepts of created things, Leibniz doubted that existence could be seen as one of the predicates forming the individual's essence. Rather, existence, he says, is what is demanded by the individual's essence (more on this below). If a given essence has a certain claim for existence, this implies that it can exist, but also can *not* exist. If so, Leibniz does not regard 'existence' as one of the predicates that make up the individual's complete concept. Leibniz, however, did not generalize this view; rather, he made a very significant exception (which Kant explicitly attacks). According to Leibniz, there is a unique and necessary being whose essence *does* include existence as one of its essential perfections, namely, God. Along with a long tradition, Leibniz argues that God is the most perfect being whose essence includes existence as one of his perfections.[11]

What I shall attempt to highlight here is that Leibniz's approach to the question of existence is closely related to his view of possibility. Leibniz developed a conception of possibility in which the traditional notion of essence is understood in terms of conceptual self-consistency. More precisely, for Leibniz, conceptual self-consistency serves not merely as a necessary condition for possibility but also as a sufficient condition for possibility. In some places, Leibniz goes as far as arguing that even the essence of God must be shown to be possible (i.e., to be self-consistent).

completely formed in his understanding' (LR 109)] must be complete and determined' (LR 108). The notion of priority here is, of course, logical, not temporal.

[10] It is arguable that 'being chosen by God' or 'being part of the best possible world' or 'being such that God will choose to actualize it' would be included in the Leibnizian concept of an individual. In this case, something that would lead to existence might be included in the concepts of these individuals. I develop this option in the next and last sections.

[11] For example, A 6.4 18–19, (1677); A 2.1 390–3, (1678); G IV 405–6, (1701).

He nevertheless adheres to the traditional view according to which existence belongs to God essentially, so that essence and existence are inseparable in God alone.

Leibniz's adherence to this privileged status of God's existence has led Russell to claim that Leibniz equivocates on the notion of existence, namely, claiming that it is a predicate in the case of God (which he sees as a necessary being) but not as a predicate in the case of creatures (which he sees as contingent beings).[12] I think that Russell was right on this score. I believe that Leibniz's equivocation brings out one of his deep metaphysical commitments, showing why he was not prepared to generalize this point, and why Kant ultimately was prepared to do so.[13] This will help in explicating Leibniz's reasons for distinguishing between two notions of existence, which apply systematically to two kinds of things—created and non-created. While my aim here is not to defend Leibniz, I will show that his equivocal notion of existence is a systematic, and well-motivated distinction between the kind of existence that pertains to the necessary being, on the one hand, and the kind of existence that pertains to contingent (created) things, on the other.

Given this background, Kant's point that existence is not a predicate is better seen as a generalization of Leibniz's line of reasoning regarding created things (extending it to the concept of God) rather than as a novel point. The full story, however, is far more complicated and its telling requires some complex historical and philosophical context. This chapter attempts to make a modest contribution to the reconstruction of this complex context. I stress that it is modest because there is much that this chapter leaves out. I do not address in any detail the question of Leibniz's actual influence on Kant. In particular, I disregard almost entirely the details of the complex way in which Kant received Leibniz's views,[14] as well as other important sources for Kant's views on existence and possibility, such as Wolff, Baumgarten, and Crusius.[15]

As will become apparent, I am focusing here on philosophical change in the relations between the notions of existence and possibility in Leibniz and Kant. The story I tell begins with Leibniz's formulation of a strictly logical notion of possibility in his early Paris notes, proceeds with Kant's pre-critical statement in 1763 that existence is not a predicate, and ends with the *Critique of Pure Reason*.

[12] Bertrand Russell, *A Critical Exposition of the Philosophy of Leibniz* (London: Allen and Unwin, 1937), p. 185. For more on the significance of Russell's reading of Leibniz for his thinking about modality, see chapter 6 of the present volume.

[13] Leibniz's equivocation might be partially explained by reference to the development of his views so that the view of existence as a predicate is more evident in the early period (roughly up to the later 1670s). This line of interpretation has been suggested by Adams, *Leibniz: Determinist, Theist, Idealist* (New York: Oxford University Press, 1994) and followed recently with more details on Leibniz's development by Mogens Lærke, *Leibniz lecteur de Spinoza* (Paris: Honoré Champion, 2008), sect. III, 2.3. I address this suggestion in the next section.

[14] In particular, I do not treat the important and complicated question of how distinct were Leibniz's views from the way in which Wolff has understood and presented them.

[15] See, for instance, Toni Kannisto, 'Positio contra complementum possibilitatis—Kant and Baumgarten on Existence', *Kant-Studien* 107/2 (2016): 291–313.

The second section presents Leibniz's view of possibility against the traditional view of temporal (or statistical) modality; the third section presents his (twofold) notion of existence. The fourth section considers Kant's pre-critical essay of 1763 and argues that Kant's view that existence is not a predicate is strongly related to the logical view of possibility advanced by Leibniz. The fifth section looks at Kant's transition to the critical period and its implications on the analysis of modality. I conclude with a rough sketch of the relations between possibility and existence between Leibniz and Kant.

## 3.1 Leibniz on Possibility

An adequate presentation of Leibniz's view of possibility would require a book. The gist of Leibniz's view of modality may be presented succinctly as a set of presuppositions he held from his early to his late writings. In *Possibility, Agency, and Individuality in Leibniz's Metaphysics,*[16] I presented these presuppositions in some detail and argued that they add up to a combinatorial approach to possibility. Here I abbreviate this account with a view to the major developments that take place in the analysis of modality between Leibniz and Kant. The main features of Leibniz's approach to modalities can be summarized as follows: in this approach, the notion of possibility is explicated in terms of consistent combinations of unique elements or terms. In other words, possibilities are seen as consistent relations among terms and pertain primarily to concepts or thoughts, not things.

Leibniz's view of possibility is situated in a conceptualist/mental framework. Possibilities are seen as thoughts of God's intellect, not as entities or as potential states of existing things. This situates Leibniz's position with respect to two important traditions: as a conceptualist position with regard to the debates between realists and nominalists concerning the status of *possibilia* and relations among the late scholastics; and in contrast to the Aristotelian account of possibility, which is grounded in the notion of potentiality. In contrast to the Aristotelian notion of possibility, according to which all genuine possibilities—seen as potential states of things—will be realized, Leibniz understands possibilities as consistent combinations of terms with no reference to existing things. In Leibniz's approach, something is possible if its concept is free of contradictions, impossible if its concept involves a contradiction, necessary if its opposite involves a contradiction. As he writes, 'all truths that concern possibles or essences and the impossibility of a thing or its necessity (that is, the impossibility of its contrary) rest on the principle of contradiction; all truths concerning contingent things rest on the principle of perfection' (AG 10). For Leibniz, what's possible is defined by the principle of contradiction (and is the realm of pure logic) and what's to become

[16] Ohad Nachtomy, *Possibility, Agency, and Individuality in Leibniz's Metaphysics* (Dordrecht: Springer, 2007).

actual (among possibles) is decided according to the principle of sufficient reason (and is thus contingent upon God's choice to realize it).

Leibniz's approach to possibility is also actualist in the sense of presupposing something actual as its grounds. In particular, Leibniz is presupposing God's mind and his simple attributes as the actual basis of producing (or thinking) possibilities.[17] In this framework, possibilities are understood as God's thinking all the combinations among his simple attributes or forms. I call this a combinatorial approach to possibility. It is logical rather than temporal, conceptualist rather than realist or nominalist, and actualist, in the sense explained above. In order to substantiate these general features let us take a closer look at some of Leibniz's presuppositions.

### 3.1.1 *Possibilities as thoughts*

Leibniz's first supposition is that *possibilities are conceived in God's mind.* This supposition mainly concerns the status of possibilities (that is, the notion of *possibilitas,* in the scholastic jargon) rather than the question what type of things are considered to be possible (the question of *possibilia,* in the scholastic jargon). Looking at this supposition will also help us to place Leibniz's view of possibility in its historical context. Leibniz's view can be situated within the tradition stemming from Scotus's logical interpretation of modal notions.[18] However, Leibniz did not just follow this tradition; he also radicalized and systematized it. For Leibniz, the ontological background of possibilities, seen as platonic forms rendered as the essence of a Christian God, was not only unnecessary, as it was for Scotus, but was also misleading. According to Leibniz, the platonic realm of essences and intelligible entities becomes a realm of pure logical possibilities. This subtle and seemingly innocuous change has dramatic consequences for the status of possibilities. In fact, it signifies a crucial turn in the history of the notion of possibility. Possibilities need no longer be seen as entities subsisting in God; they need no longer be seen as some type of shadowy entities at all. Rather, Leibniz does not see possibilities as entities in the first place but as mere thoughts in God's understanding.[19] With this deflation, the very notion of intelligibility is transformed as well: from its platonic sense of true Being to that which can be *understood* by a perfect mind—regardless of whether it exists or not.

Seeing possibilities as thoughts is very significant. For Leibniz, possibilities are pure essences that may or may not exist. This Leibnizian notion of essence is obviously different from a platonic-like essence. An essence, for Leibniz, is a cluster

[17] For further clarification of this sense of actualism, see Adams, *Leibniz: Determinist, Theist, Idealist.*

[18] See S. Knuuttila, 'Modal Logic' in *The Cambridge History of Later Medieval Philosophy*, edited by N. Kretzmann, A. Kenny and J. Pinborg (Cambridge: Cambridge University Press, 1982), 342–57, p. 344.

[19] See Mondadori, 'Modalities, Representations, and Examplars: The "Region of Ideas"', in *Mathesis Rationis*, edited by A. Heinkamp, W. Lenzen, and M. Schneideger (Münster: Dutz, 1990), 169–88, and M. Mugnai, *Leibniz's Theory of Relations, Studia Leibnitiana Supplementa* 28 (Stuttgart: Franz Steiner, 1992) for more details and some substantiation of this claim.

of predicates that specify the properties of a thing, which may or may not exist. In this way, the traditional distinction between essence and existence becomes a real distinction. Simply put, possibilities do not exist; they are merely thought or conceived by God. In other words, possibilities do not have mind-independent existence; they exist merely as thoughts or conceptions of God. To be clear, possibilities do not have some particular form of being which is not existence.

And it is important for Leibniz that some of the possibles shall never be realized, but at the same time serve as a precondition for worldly existence. In contrast to the platonic interpretation, for Leibniz, truths need not be seen as mind-independent entities, or as entities of any kind. It suffices that they are conceived or thought in God's mind. Indeed, Leibniz ascribes a similar conceptual status to truths, numbers and relations.[20] The significance of separating the realm of truth and possibility from that of reality cannot, in my mind, be overstated. It produces a conceptual separation between the thinkable (i.e., the possible), and the real. Likewise, this distinction corresponds to a distinction Leibniz draws between two notions of possibility and impossibility: 'one from essence, the other from existence or, positing as actual' (A 464; DSR 7).

In addition, Leibniz's logical interpretation of possibilities implies a dramatic rejection of the dominant interpretation of modal terms during the early modern era, namely, the Aristotelian—temporal or statistical—interpretation of modalities. In fact, the true novelty of Leibniz's view of possibility can be grasped only in contrast to this view of possibility. According to the Aristotelian view, possibilities roughly correspond to the potential states of existing things,[21] so that a state of affairs is possible if it either occurs in the present, will occur in the future or has occurred in the past. If a state of affairs has not occurred, does not occur, and will not occur, then, it is considered impossible.[22]

Following Knuuttila, it is instructive to characterize the Aristotelian view according to the principle: '*no genuine possibility can remain unrealized*'.[23] In this view, the modal terms (i.e., possible, impossible, necessary and contingent) are defined in direct reference to time. For example, if there is a moment in time in which a statement is true, then it is possible; if a statement is false at all times, then it is impossible; if a statement is true at all times, then it is necessary; if a statement is true at some time and false at another, then it is contingent.[24]

[20] 'Numbers, modes, and relations are not entities' (A 463; DSR 7).

[21] This view was articulated by Aristotle and I label it 'Aristotelian', following Hintikka and Knuuttila. However, in Aristotle's writings, one finds other views as well. Roughly speaking, the view that any genuine possibility will be actualized was also held at least by Spinoza and Hobbes. Regarding this point, see the first section of Hintikka's 'Leibniz on Plenitude, Relations, and the "Reign of Law"' in *Reforging the Great Chain of Being*, edited by S. Knuuttila (Dordrecht: Springer, 1981), 259–86.

[22] See Aristotle's *Metaphysics*, Theta 4; 47b3.

[23] For the formulation and illuminating discussion of this principle, see S. Knuuttila, 'Modal Logic'.

[24] See Knuuttila, 'Modal Logic', p. 344.

Leibniz's denial of the temporal interpretation has an additional significant implication, viz., a denial of the principle that every possibility will be realized.[25] Leibniz's rejection of the principle that *no genuine possibility can remain unrealized* implies disagreement with most of his contemporaries, including Hobbes, Descartes, and Spinoza.[26] From this partial list it can be seen that, although the logical view of possibility has almost become common sense to us today, in the early modern period, it was certainly a minority view—especially among the great modern philosophers.

### *3.1.2 Possibilities as consistent thoughts*

The next supposition is familiar and seemingly innocuous, namely, that possibilities correspond to consistent thoughts.[27] This is the heart of the logical interpretation of possibility. We have already seen that Leibniz defines possibilities in terms of divine thinkability and intelligibility. The supposition of self-consistency provides the crucial constraint on this notion of intelligibility. The notion of self-consistency clearly belongs in a conceptual realm, for it presupposes consistency among *terms*. A contradiction cannot arise among entities, only among terms.[28] However familiar, the analysis of self-consistency in the Leibnizian context reveals some less familiar presuppositions.

For one thing, this supposition shows that the notions of thinking and possibility are intrinsically connected.[29] Leibniz defines genuine thoughts in terms of possible, i.e., self-consistent, concepts and defines possibilities in terms of the intelligible activity in the divine understanding. A clear statement of this view appears as early as the *Confessio philosophi* of 1673, in which Leibniz writes:

> I have defined the necessary as that the contrary of which cannot be understood; therefore, necessity and impossibility of things are to be sought in the ideas of things themselves, and not outside those things, by examining whether they can be thought or whether they imply a contradiction. (A 128; CP 57)

[25] For obvious reasons, the principle that any genuine possibility will be realized has been viewed (by Knuuttila et al.) as the adequate rendering of the principle of plentitude. Leibniz's own version of the principle of plentitude will be qualified, so that only compossible individuals are seen as candidates for creation while there remain infinitely many unrealized logical possibilities.

[26] 'If everything that exists were necessary, then it would follow that only things which existed at some times would be possible (as Hobbes and Spinoza hold) and that matter would receive all possible forms (as Descartes held). And so, one could not imagine a novel that did not actually take place at some time and in some place, which is absurd. And so, we should say rather, that from an infinite number of possible series, God chose one for reasons that go beyond the comprehension of his creatures' (The Source of Contingent Truths, AG 100).

[27] In Leibniz's view, there are no inconsistent divine thoughts, only human ones (which are seen as inconsistent concatenations of terms).

[28] It is worth noting that Kant emphasizes that a real repugnance can exist between entities (even though this is not strictly speaking a contradiction, since it involves entities rather than terms). See for instance the Negative Magnitudes essay. This pertains to Kant's notion of real possibility, which I discuss in section 3.4. I thank Reed Winegar for this note.

[29] In the De Summa Rerum he also says that, 'every thing possible is thinkable' (A 475; DSR 27–9).

Leibniz goes on to define the modal notions as follows:

> I call that necessary, the opposite of which implies a contradiction, that is, that which cannot be clearly understood.... Those things are contingent that are not necessary; those are possible whose non-existence is not necessary. Those are impossible that are not possible, or more briefly: the possible is *what* can be conceived, that is (in order that the word 'can' does not occur in the definition of possible) *what is understood clearly by an attentive mind*; the impossible—what is not possible. (A 127; CP 55)[30]

Observe the subtle shift from 'what can be conceived' to 'what is understood by an attentive mind'. This indicates an actualist strand in Leibniz's theory of possibility since the notion of pure logical possibility is grounded in the actual thoughts of God. Because Leibniz identifies the possible with what is conceived by God's understanding, he also considers the essence of a given thing as independent of its existence. As we shall see, he does this by interpreting the essence of a thing as the complete concept or the possibility of that thing.

> If the essence of a thing can be conceived... (e.g., a species of animal unequally footed, also a species of immortal animals) then surely it must be held to be possible, and its contrary will not be necessary, even if perhaps its existence is contrary to the harmony of things and the existence of God, and consequently will never exist.... Hence all those who call impossible (absolutely, i.e., *per se*) whatever was not nor is nor will be are mistaken. (A 128; CP 57)

Leibniz alludes here to the traditional Aristotelian sense of possibility that is grounded in the past or future existence of things. By contrast, his clear distinction between essence and existence gives rise to two distinct notions of possibility. As he writes in his Paris notes:

> Impossible is a two-fold concept: that which does not have essence, and that which does not have existence, i.e., that which neither was, is, nor will be because it is incompatible with God, or with the existence or reason which brings it about that things exist rather than do not exist. One must see if there are essences which lack existence, so that it cannot be said that nothing can be conceived which will not exist at some time in the whole of eternity.

His response to his own query is clear: 'The origin of impossibility is twofold: one from essence, the other from existence or, positing as actual' (A 464; DSR 7). There are many genuine possibilities, i.e., logically possible things, that will never come to exist, as there are many possible situations that are intelligible but never occur in our world. Within the realm of the conceivable, many creatures and things that do not exist may be conceived.

> This is the mark of an elegant poet that he fabricates something false but nevertheless possible. The *Argenis* of Barclay is possible i.e., is clearly and distinctly imaginable.... The *Argenis* would

[30] The passage continues as follows: 'the necessary—that whose opposite is impossible; the contingent—that whose opposite is possible.' See also On Freedom and Possibility (1680–82): 'all truths that concern possibles or essences and the impossibility of a thing or its necessity rest on the principle of contradiction' (AG 19).

not have been impossible, although it did not yet exist. Those who think otherwise necessarily destroy the difference between truth and possibility, necessity and contingency

(A 128–9; CP 57–9).

Similarly Leibniz notes: 'when we dream of palaces, we rightly deny that they exist' (A 464; DSR 7). Any thought that is conceivable (non-contradictory) is logically possible in itself, but not all such possibilities are realized in the created world. Thus a distinction is drawn between truth and possibility. Consequently, there are many statements (e.g., hypothetical ones) that need not refer to existing things at all. One can speak of the properties of the Argenis of Barclay without being committed to the Argenis' existence, given that he is logically conceivable. In other words, there is a concept of the Argenis just as there are concepts of the palaces in the *Arabian Nights*. A fictional character, fabricated by an elegant poet, is not impossible—even if it does not (and will never) exist in our world. As we know, this point plays a central role in Leibniz's metaphysics. For instance, his rejection of absolute necessity and his theodicy project of justifying the goodness of the actual world depend on it.[31]

When Leibniz defines his modal notions in terms of intelligibility or conceivability, it is not human capacities that he primarily has in mind but God's.[32] 'God is that which perceives perfectly whatever can be perceived' (A 519; DSR 79). Since God is all-knowing and his mind is infinite, any conceivable thing will be conceived by him. Humans are limited in their capacities to think and understand ideas. For example, it requires some intellectual work for humans to realize that 'the number of all numbers' does not express a genuine notion. Leibniz believed that the methods of analysis and synthesis enable and facilitate such intellectual work for humans. God's omniscient intellect, however, is not limited in this way: God sees at once that the combination of ideas, 'the "number" of all "numbers"', implies a contradiction. Therefore it cannot be 'distinctly conceived'.[33] Thus, there is no notion of it or its notion is impossible (see, for example, A 463; DSR 7, A 520; DSR 79).

### *3.1.3 Consistent thoughts, complex concepts and simple constituents*

As it turns out, consistency does not apply in the limit case, that is, in the case of God's most simple forms. Leibniz presupposes that complex concepts are composed of simple ones. In his Paris notes Leibniz remarks that 'there are necessarily simple forms' (A 514); and that 'nothing can be said of forms on account of their simplicity' (A 514; DSR 69); such forms are unanalysable and indefinable (A 572). However,

[31] See On Freedom, in AG 94.

[32] These definitions apply to humans as well, though in a limited sense. This is why humans can benefit from philosophical and logical investigations and why they should devise the combinatorial and logical calculi. In my view, the projects of the universal language and the real characteristic are premised on this presupposition. They present human attempts to model and investigate divine thoughts by using symbolical and linguistic representation.

[33] See A 583; DSR 105–7. It seems that, for Leibniz, that which can be 'distinctly conceived' is that which can be conceived by considering the concepts, regardless of their instantiation in the world.

owing to their simplicity, the relation of consistency among terms cannot apply to simple forms; it can only apply to complex concepts. A simple constituent 'A' cannot be either consistent or inconsistent simply because it is not complex. If it has no constituents, there are no relations among its constituents, that it can be regarded as either consistent or inconsistent. This observation points to another Leibnizian presupposition.

If possibility is to be ascribed only to complex concepts whose constituents are self-consistent, then it may only be ascribed to complex concepts. The compatibility relations between the terms (or constituents) can only apply to concepts composed of certain terms, such as 'the' 'greatest' 'number'; 'the' 'most' 'rapid' 'motion'; 'a' 'winged' 'horse', but it cannot apply to absolutely simple or atomic concepts. For this reason we may say that Leibniz's approach to possibility presupposes compatibility (and incompatibility) relations among the constituents of complex concepts (their terms in the traditional jargon), as well as their co-consideration in God's mind. The simplest concepts, however, cannot be seen as possible in the same sense as the complex, since they do not satisfy a necessary condition for logical possibility, that is, self-consistency. Instead, God's simple forms may be seen, in accordance with Leibniz's theological commitments, as actual. Indeed the presupposition of God's mind and its simple forms are part of the actualist basis for Leibniz's theory of possibility.

The following passage from the *Combinatorial Art* reveals some of Leibniz's early presuppositions regarding the compositional nature of concepts:

> Since all things which exist, or which can be thought of are in the main composed of parts, either real or at any rate conceptual, it is necessary that those things which differ in species differ either in that they have different parts – and here is the use of complexions – or in that they have a different situation – and here is the use of dispositions. The former are judged by the diversity of matter; the latter, by the diversity of form. With the aid of complexions, indeed, we may discover not only the species of things but also their attributes. Thus almost the whole of the *inventive* part of logic is grounded in complexions – both that which concerns simple terms and that which concerns complex ones. (L 130)

The compositional structure of concepts motivates and informs Leibniz's enterprise to discover and analyse all complex concepts. In turn, the discovery of all possibilities and impossibilities motivates the analysis of concepts as well as the discovery of new concepts. Donald Rutherford nicely brings out Leibniz's supposition of the combinatorial nature of concepts here:

> In [the *Combinatorial Art*] we meet full-blown the theory of the combinatorial nature of concepts—the doctrine that all complex concepts are composed from, and analyzable into, simpler concepts—a constant feature of all of Leibniz's later writings. It is evident that he regards this theory as following from more general metaphysical principles. In his view, *all* things and thus all concepts, are defined in terms of the parts they contain (their 'matter') and the specific arrangements of these parts (their 'form'). Differences of parts... are

differences of 'complexion'; differences in the arrangement of parts are differences of 'situation' or 'disposition.'[34]

Leibniz's notion of 'matter' refers to the elements that constitute a complex concept while his notion of 'form' refers to the various ways in which the elements (i.e., matter) may be arranged or ordered. Leibniz's master insight into the notion of possibility, however naïve it might be, can be stated in these terms: Given a number of simple concepts, call them 'matter', possibilities may be accounted for by virtue of variations in form.[35]

### *3.1.4* Prima possibilitas *and God's attributes*

This presupposition relates to another: Leibniz identifies the simple conceptual elements, which he calls simple forms or perfections (A 575, 578), with the simple attributes of God (an identification goes back to Lull's *Ars magna*). He writes that, 'God is the subject of all absolute simple forms' (A 519; DSR 79) and adds that, '[a]n attribute of God is any simple form' (A 514; DSR 69). This supposition provides the basic assumption for his proofs that a most perfect being is possible and consequently that a most perfect being exists (A 572–79).[36] Leibniz's proof that all positive perfections that belong to God's concept are compatible (*inter se*) is grounded in the supposition that these perfections are simple (and positive).

God's simple forms may be seen as the material or the actual basis out of which possibilities arise in God's mind by virtue of mental combinations and reflections. Leibniz's supposition of simple forms as the 'elements of thinking' suggests that his combinatorial approach to the construction of concepts and possibilities is recursive, that is, it has simple forms as its ultimate starting point.

This supposition of 'logical atomism' (if I may use this anachronistic term) plays an important role in Leibniz's approach to possibility: (1) The presupposition of absolute simple forms accords with Leibniz's notion of a natural order, that is, proceeding from the simple to the complex in the construction of possibilities. (2) The postulation of different simple forms would allow Leibniz to account for negations and the incompatibility relations among predicates of complex concepts.[37]

[34] D. Rutherford, 'Language and Philosophy in Leibniz', in *The Cambridge Companion to Leibniz*, edited by N. Jolley (New York: Cambridge University Press, 1995), pp. 227–8.

[35] In chapter 2 of my *Possibility, Agency, and Individuality in Leibniz's Metaphysics*, I argue that the significance of Leibniz's notion of form (the order and arrangement of the elements) for his notion of possible individuals has been largely underestimated by Leibniz's commentators.

[36] 'Attributum Dei est, forma simplex quaelibet' (A 514). See also A 522. In the context of proving that 'a most perfect being exists', which he defines as 'a being which is the subject of all perfections' (A 577; DSR 101), he writes: 'Perfections, or simple forms, or absolute positive qualities, are indefinable or unanalyzable' (A 575; DSR 97). 'I term a "perfection" every simple quality which is positive and absolute, or, which expresses without any limits whatever it does express. But since a quality of this kind is simple, it is therefore indefinable or unanalyzable' (A 577; DSR 99). See also A 578.

[37] 'Every purely affirmative attribute is infinite; or, it is great as it can be, or contains all things that belong to its genus. There are necessarily several affirmative primary attributes; for if there were only one,

Indeed, I think that this is also the ultimate source of incompossibility relations among possible individuals.[38]

The identification of simple forms or the *prima possibilitas* with the perfections or the simple attributes of God, defined as 'the subject of all absolute and positive perfections' (A 577; DSR 101), is intrinsically related to the metaphysical and theological contexts of Leibniz's approach to possibilities. Leibniz sees possibilities as self-consistent thoughts in God's omniscient mind. Given Leibniz's presuppositions presented above, we may interpret this point as follows: seen as the 'subject of all perfections' and as an active (thinking) agent,[39] God thinks the various combinations among his simple forms so that complex concepts or possibilities arise in his mind (see A 514; DSR 71). This implies that God combines the simple forms in a natural order—from the simple to the complex. It is in this rich sense that possibilities are conceived in God's understanding.

## 3.2 Leibniz on Existence

> 'Existence does not differ from Essence in God, or, what is the same thing, it is essential for God to exist. Whence God is a necessary being.
>
> Creatures are contingent, that is, their existence does not follow from their essence.'
>
> (On Contingency, AG 28)

In the context of Leibniz's attempts to prove God's existence (in early 1676), we find him employing two distinct notions of existence. In his modified version of Anselm's proof, formulated in two (tightly related) notes *Quod Ens Perfectissimum Sit Possible* (That the Most Perfect Being is Possible) (A 6.3 572–74) and *Quod Ens Perfectissimum Existit* (That the Most Perfect Being Exists) (A 6.3 578–80), Leibniz considers existence to be a perfection. According to Leibniz, the validity of Anselm's argument (as revived by Descartes) depends on showing that the notion of the most perfect being is consistent. As he writes, 'there is given, or, there can be understood, a being which is the subject of all perfections, or a most perfect being. Hence, it is at once evident that it exists; for existence is contained among perfections (*cum et existentia inter perfectiones contineatur*)' (A 6.3 577; Pk 101).

According to Leibniz in these notes, if the concept of the *Ens Perfectissimum* is shown to be possible, one would immediately see that such a being necessarily exists because existence (seen as a predicate, perfection, or an attribute) belongs to its

only one thing could be understood. It seems that negative affections can arise only from a plurality of affirmative attributes—for example, thought and extension' (A 572–73; DSR 93).

[38] I develop this in 'On The Source of Incompossibility in Leibniz's Paris Notes and some Remarks on Time and Space as Packing Constraints', in *Compossibility and Possible Worlds*, edited by Yual Chiek and Gregory Brown (Dordrecht: Springer, 2015).

[39] 'There is a uniquely active thing, namely, God' (A VI, ii, 489).

essence. In these texts, Leibniz defines '*Ens Perfectissimum*' as the subject of all perfections or of all absolute positive attributes. As Fichant pointed out, the compatibility among perfections here means 'the possibility of the *in esse* of different predicates in the same subject'.[40] Once this definition is shown to be consistent, the proof proceeds on the familiar Anselmian and Cartesian assumption that existence is a perfection and thus would be included in a subject that contains all perfections.[41] Leibniz also defines God as 'a being from whose possibility (or from whose essence) his existence follows' (A 6.3 582). Leibniz's adherence to the validity of this argument shows that he presupposes that existence is a constitutive predicate of the concept of the most perfect being.[42] As he writes:

> Again, a necessary being is the same as a being from whose essence existence follows. For a necessary being is one which necessarily exists, such that for it not to exist would imply a contradiction, and so would conflict with the concept or essence of this being. And so existence belongs to its concept or essence. (A 6.3 583)

Leibniz was quite proud of his 'possibility proof' (or his amendment to Descartes' proof) in 1676. He showed it to Spinoza and noted with evident pride that, after some explanation, Spinoza had approved of it (A 579). As far as we know, Leibniz never renounced this version of his argument. He mentions his original contribution to this proof throughout his career and hardly misses an opportunity to attack the Cartesians for adhering to a non-valid form of it.[43]

Leibniz's demand that the notion of God be shown to be consistent as a precondition for establishing God's existence should be seen as part of a more general and novel approach regarding the relation between possibility and existence. For Leibniz, existence claims (such as 'God exists') presuppose possibility claims or what he calls real definitions (that the definition of God is consistent). According to Leibniz, that something *can* exist is logically prior to whether it exists or does not exist. To show that something is possible requires showing that its concept is consistent. This is the point of giving a real definition—a definition establishing the consistency of a given concept.

Yet, as Russell had already noted in 1937, Leibniz's position regarding existence is more complex.[44] In addition to thinking of existence as one of God's perfections, Leibniz also employs a different notion of existence. This is already apparent in the

[40] M. Fichant, 'L'origine de la négation', in Fichant M., *Science et métaphysique dans Descartes et Leibniz* (Paris: Press Universitaires de France, 1998), p. 111.

[41] For a critique of this argument, see Mogens Laerke, *Leibniz lecteur de Spinoza, La genèse d'une composition complexe* (Paris: Honoré Champion, 2008) sect. IV, 6.2.

[42] For this reason, it is misguided to play down Leibniz's view of existence as a perfection, as J. J. Vilmer does in 'L'existence leibnizienne', *Archives de philosophie* 70/2 (2007): 249–72. Vilmer (p. 255) states emphatically against Russell and Mates that 'l'existence pour Leibniz n'est pas une perfection'. Since Leibniz makes it explicit that he amends and modifies Descartes' proof by showing that the notion of a most perfect being is consistent, this judgement seems to be off the mark.

[43] To mention a few notable examples: Letter to Countess Elizabeth, 1678 (AG 240); *Meditations on Knowledge, Truth, and Ideas*, 1684 (AG 25–26).

[44] See Adams, *Leibniz: Determinist, Theist, Idealist* for more details.

same set of texts—the Paris notes of 1672–76. For example, in the *De Arcanis Sublimium Vel De Summa Rerum,* from February 1676, Leibniz writes: 'to exist is nothing other than to be harmonious (*Existere nihil aliud esse quam Harmonicum esse*)' (A 6.3, 474; Pk 24–5). In the *De Existentia* from late 1676 Leibniz notes: '[...] for things to exist is the same as for them to be understood by God to be the best (*res existere idem est, quod a Deo intelligi optimas*).'[45] In 1678 Leibniz defines existence as that which is 'compossible with the most perfect'.[46] He also claims that '[e]xistent is the series that involves more of reality'.[47] The notion of existence sketched in these passages is obviously different from the one noted above, that is, being a simple perfection or a predicate. Among other things, these passages clearly imply that, in this context, existence is seen as a certain *relation* among possible things—the most harmonious set, the best, the most perfect, or the one with the most reality. Existence in this sense is a relation involving compossibility and the highest perfection among a subset of possible things, rather than a monadic (one place) predicate of an individual.

In Leibniz's metaphysics, such a relation arises in a well-defined context, namely, as God conceives and compares all possibilities as candidates for creation.[48] Note that, in this context, the existence of creatures *presupposes* the existence of God as the perceiver of all possible individuals and their relations. As Leibniz notes explicitly (between 1675 and 1676): 'I seem to have discovered that to Exist is nothing other than to be sensed—to be sensed however, if not by us, then at least by the Author of things, to be sensed by whom is nothing other than to please him or to be Harmonious' (A 6.3 56). It seems clear that Leibniz's conception of existence here is richer than being considered as a mere perfection or an attribute of an individual thing. In *Existentia. An sit perfecto* of 1677, Leibniz states flatly that existence is not a perfection: 'In effect, it is true that that which exists is more perfect than the non-existing, but it is not true that existence itself is a perfection, for it is nothing but a certain comparison between perfections [*perfectionem inter se comparatio*]' (A 6.4, 1354).[49] Existence is a specific *relation* among possible individuals, viz., the one that picks out the best or the most harmonious set of individuals.

Robert Adams has argued convincingly that this latter definition of existence should be understood in Leibniz's sense of a real definition, that is, as an explication of the nature of existence, so that it would 'exhibit the reason or cause that an existence would have.'[50] Such an explication would run along the following lines: since the existing set of possible things was perceived by God to be the most perfect

[45] A 6.3, 588; DSR 30. [46] See also A 6.4 867.

[47] See also A 6.4 2770 and Adams, *Leibniz: Determinist, Theist, Idealist*, p. 168.

[48] See Mugnai, 'Leibniz's Nominalism and the Reality of Ideas in the Mind of God', in *Mathesis rationis*, edited by A. Heinekamp, W. Lenzen, and M. Schneider (Münster: Dutz, 1990); B. Mates, *The Philosophy of Leibniz: Metaphysics and Language* (New York: Oxford University Press, 1986); and chapter 3 of my *Possibility, Agency, and Individuality in Leibniz's Metaphysics*.

[49] Cf. A 6.4, 1445.

[50] Adams, *Leibniz: Determinist, Theist, Idealist*. It is worth noting that Leibniz develops his notion of real definition in the very context of proving that the notion of '*Ens Perfectissimum*' is possible.

and harmonious, it was selected for actualization. In this sense, this definition gives the reason or cause for the existence of created things, that is, the actual world. But we must not overlook the role of God as 'the reason or cause that an existence would have' in this reasoning. As Leibniz would write later: 'If there were no necessary being, there would be no contingent being either. For a reason must be provided why contingents should exist rather than not exist' (A 6.4 1617).

No matter how exactly we construe this second notion of existence, it clearly presupposes God's existence. If 'to Exist is nothing other than to be sensed ( ... ) not by us, [but] by the Author of things', then God's existence clearly figures in this definition of existence. Since God's existence is presupposed in this formulation, it is very clear that this notion of existence does not apply to God's existence but only to the existence of created things. Hence it seems that Leibniz's metaphysics utilizes two senses of existence: one applicable to created (and contingent) things and another applicable to their Creator (seen as a necessary being).

This approach implies that existence is not said of God in the same way as it is said of his creatures. Leibniz avoids Russell's dilemma[51] by employing two different senses of existence.[52] While Leibniz may be accused of equivocation, I do not see a compelling reason to think that such a dual sense of existence would be unwelcome to him.[53] The gap between the Creator and his creatures would seem substantive enough to permit a systematic distinction in the application of notion of existence. To a large extent, the distinction between the eternal and necessary existence of the Creator and the temporal, contingent, and dependent existence of creatures is part of a theological tradition that goes a long way from Augustine to Spinoza and Descartes.

While Leibniz's conviction in his a priori proof for the existence of God waned over the years, his theory of creation never disappears; rather, it persists throughout his career. Thus, if, as Russell argues, something has to be given up by Leibniz, the texts strongly suggest that it would be the a priori proof for God's existence in favour of a mere presumption that God is possible, rather than his theory of creation.[54]

[51] Russell argued that, if existence is seen as a predicate, Leibniz's theory of creation has to be given up (for then the actual world would exist by definition and there would be no need for an act of creation); but if existence is not a predicate, the ontological argument would have to be given up: 'either creation is self-contradictory, or, if existence is not a predicate, the ontological argument is unsound' (Russell, *A Critical Exposition of the Philosophy of Leibniz*, p. 185).

[52] 'Except for the existence of God alone, all existences are contingent. Moreover, the reason [*causa*] why some particular contingent things exist, rather than others, should be sought not in its definition alone, but in a comparison with other things. For, since there are an infinity of possible things which nevertheless, do not exist, the reason [*ratio*] why these exist rather than those should not be sought in their definition (for then nonexistence would imply a contradiction, and those others would not be possible, contrary to our hypothesis) but from an extrinsic source, namely, from the fact that the ones that do exist are more perfect than others.' (AG 19. See also p. 20.)

[53] Cf. the opening lines of On Contingency (circa 1986), Grua 302, AG 28; On Freedom and Possibility (1680–82) AG 19.

[54] See M. Antognazza, 'Arguments for the Existence of God: The Continental European Debate', in *The Cambridge History of Eighteenth-Century Philosophy*, edited by K. Haakonssen (Cambridge: Cambridge University Press, 2006), 731–48, p. 736.

I would suggest that Leibniz in fact held on to both positions by systematically using two notions of existence—one sense applicable to the Creator, and another applicable to creatures. I do not see much textual evidence to think otherwise.[55]

By way of recapitulation, let us note that Leibniz clearly articulates the idea that existence does not add anything to a concept of an individual. The idea that existence is not a predicate is applicable to created things but not to God. In addition, Leibniz holds that possibility—defined as self-consistency among terms—presupposes something actual, namely, God's attributes and understanding. Leibniz conceives of God as the mind who thinks all possibilities and whose attributes constitute an actual ground for thinking possibilities.

For Leibniz, the notion of God as the grounds for all possibilities occupies a unique and indispensable place. As he puts this in the *Monadologie* 45, 'God alone (or the necessary Being) has this privilege that it must exist, if it is possible'.[56] This is what Leibniz calls the 'pinnacle of modal theory'.[57] The result is that God's existence, as the mind comprehending all possibilities, is *presupposed* (to be actual) in Leibniz's view of possibility.[58] As we shall presently see, Kant takes up a similar position in his early essay from 1763, to which I now turn.

## 3.3 Kant's Pre-Critical Period: Existence and Logical Possibility in the '*Beweisgrund*' (1763)

Kant begins his 'The Only Possible Argument in Support of a Demonstration of the Existence of God' (*Beweisgrund,* 1763) with a very clear statement that 'existence is

[55] Leibniz was clearly dissatisfied with the traditional formulation of the ontological argument and sought to support it with an a priori proof that the most perfect being is possible. In 1676 he was convinced that he could provide such a proof but it is not so clear how long his conviction lasted. As Adams notes, after 1678, the a priori possibility proof of God's existence no longer surfaces in Leibniz's texts and what becomes more prominent is a presumption in favour of the possibility of the *Ens Perfectissimum.* Leibniz's idea is that the notion of the most perfect being should be presumed possible shown to be impossible. Thus, it is arguable that Leibniz himself becomes hesitant about his early a priori possibility proof and perhaps likewise about seeing existence as a predicate (even when applied to God). Laerke has recently argued that during his correspondence with Eckhardt in 1677 Leibniz abandons his notion of existence as a perfection and *replaces* it with the notion of perfection as a degree of reality. However, I do not think that we have compelling evidence to support the claim that Leibniz *abandoned* the notion of existence as a perfection, although it clearly becomes less prominent in texts written after 1677. The notion of existence as a perfection seems to remain intact when applied to God and to coexist with the notion of existence as a degree of perfection. Laerke refers to A 2.1, 327, note 8; 329, note 3; A 2.1, 363. However, for a clear statement that Leibniz sees existence as a perfection even later, see the *Meditationes de Cognitiones, Verittate et Ideis* of 1684 GP IV 449; AG 25–26; *Monadology* § 45.

[56] *Monadology* § 44.

[57] A 6.3 583. See also *Theodicy* §184. As we shall see, for Kant in 1763, the notion of possibility presupposes a necessary being.

[58] For a substantiation of these claims, see chapters 1 and 2 of my *Possibility, Agency, and Individuality in Leibniz's Metaphysics,* as well as Robert M. Adams, 'God, Possibility, and Kant', *Faith and Philosophy* 17/4 (2000): 425–40.

not a predicate or a determination of a thing' (2:72; CE 117).[59] He goes on to justify his claim on the following grounds:

Take any subject you please, for example, Julius Caesar. Draw up a list of all the predicates which may be thought to belong to him, not excepting even those of space and time. You will quickly see that he can either exist with all these determinations, or not exist at all. The Being who gave existence to the world and to our hero within that world could know every single one of these predicates without exception, and yet still be able to regard him as a merely possible thing which, in the absence of that Being's decision to create him, would not exist (...) who can deny that in the representation which the Supreme Being has of them [all these predicates] there is not a single determination missing, although existence is not among them, for the Supreme Being cognises them only as possible things. It cannot happen, therefore, that if they were to exist they would contain an extra predicate; for, in the case of possibility of a thing in its complete determination, no predicate at all can be missing. And if it had pleased God to create a different series of things, to create a different world, that world would have existed with all the determinations, and no additional ones, which He cognises it to have, although that world was merely possible. (2:72; CE 117–18)

The Leibnizian echoes of this passage are striking. The idea that a subject is associated with an individual (a possible one in this case); the choice of the example, namely Caesar, which appears in the *Theodicy* (as well as the *Discours de métaphysique*, article 13, which Kant could not see); and that Caesar has a complete, fully determining concept, which is considered for creation as a mere possibility in 'the representation which the Supreme Being has of them', as part of a series of things (a world), are all very familiar Leibnizian themes.

In section 3 of the essay, Kant asks whether one can say that existence contains more than the possible. He responds: 'no more is posited in a real thing than in a merely possible thing, for all the determinations and predicates of the real thing are also to be found in the mere possibility of that same thing' (2:75; CE 121). He goes on to say: 'I maintain that nothing more is posited in an existing thing than is posited in a merely possible thing' (2:75; CE 121). This is the ground for Kant's claim that existence is not a predicate in this essay. Rather than a predicate, existence is 'the absolute positing of a thing' (2:73: CE 119). As he clarifies, the claim that 'X exists' does not express a relation between a subject (X) and a predicate (existence) but the (modal) position of a complete set of predicates included in the subject. While this view seems to be almost identical to the one Kant will present in the first Critique, what I would like to highlight here is the similarity between Kant's view in this essay and Leibniz's conceptualization of the relation between possibilities and the existence of created things.

[59] Thus Kant's famous point of 1781 is clearly articulated and put in general form in 1763. There is however at least a subtle terminological difference between these texts. While in 1763 the term he is using is (*Existenz* or *Dasein*), in 1781, it is not 'existence' but 'being' (*Sein*) that Kant uses in 1763. It is not clear, however, what to make of this change.

The relation between Kant's view of existence and the logical (or internal) view of possibility becomes explicit in the second Reflection. With the view of existence just presented, Kant attempts to show that the notion of internal possibility (i.e., one that does not involve a contradiction) presupposes something existing. In analysing the internal notion of possibility, he draws a distinction between formal and material elements of internal impossibility:

> ... in every possibility we must first distinguish the something which is thought, and then we must distinguish the agreement of what is thought in it with the law of contradiction. A triangle which has a right angle is in itself possible. The triangle and the right angle are the data or the material element in this possible thing. The agreement, however, of the one with the other, in accordance with the law of contradiction, is the formal element in possibility. (2:77; CE 123)

It is worth noting that, on this view, not only a possibility but also an impossibility presupposes that some things are either consistent or inconsistent. As Kant further writes, 'A quadrangular triangle is absolutely impossible. Nonetheless, the triangle is something, and so is the quadrangle' (2:77; CE 123). Kant's intuition here is that possibility as well as impossibility presuppose the things whose conjunction is either compatible or not. In both cases, the *matter* is given and presupposed: any contradiction presupposes some *things* that are contradicted. Conversely, any possibility presupposes some *things* that are in agreement with one another.

If a contradiction is a relation of opposition between two things, it clearly presupposes these things. Thus any possibility, which is defined by the principle of contradiction, requires things that are not merely possible but rather actual. In the third Reflection, section 2, entitled 'There exists an absolutely necessary being' Kant says: 'All possibility presupposes something actual in and through which all that can be thought is given. Accordingly, there is a certain reality, the cancellation of which would itself cancel all internal possibility whatever' (2:83; CE 127). Kant goes on to claim that an actual foundation of all possibility is necessary: 'it is apparent that the existence of one or more things itself lies at the foundation of all possibility, and that this existence is necessary in itself' (2:83; CE 127).[60] In subsequent sections Kant argues that such a necessary existence is unique, simple, immutable and eternal, contains supreme reality, and a mind—in short, it is something that fits all the traditional attributes of God, which satisfies Kant's declared aim here to show that such an argument provides the a priori grounds for a proof that a necessary being exists:

> The argument for the existence of God which we are presenting is based simply on the fact that something is possible... It is, indeed, an argument derived from the internal characteristic mark of absolute necessity. (2:91; CE 134–5)

The intuition behind Kant's argument can be presented thus: if something is possible, something necessary must be presupposed. Whatever subtleties and difficulties

[60] Kant defines a necessary being (along with the tradition) as that the contrary of which implies a contradiction.

Kant's reasoning in this essay may involve, the main line of argument is rather clear. The very notion of internal possibility, defined in terms of logical consistency, implies a necessary being, which Kant identifies with God. In short, the possible implies something actual, which he identifies with a necessary being, considered as the ground for all possibility.

Now, this line of reasoning is very reminiscent of Leibniz's reasoning leading to the conclusion that God is the foundation not only of all reality but of all possibility as well.[61] Leibniz's intuition can be formulated concisely thus: the notion of logical (*per se*) possibility is defined in terms of consistent thinkability, which presupposes a thinking agent and some necessary elements as the foundation of what can be thought. Early and late in his career Leibniz identifies the thinking agent with God and the simple elements with his attributes.[62]

But, unlike Leibniz, Kant does not proceed from the possibility of an *Ens Perfectissimum* to its existence but rather from the notion of possibility in general to that of an actual being. This is a significant difference between their approaches. Leibniz argues that, if the notion of a most perfect being is possible, a most perfect being necessarily exists. Kant argues that the very notion of possibility implies a necessary being. But this difference also highlights the similarity in Leibniz and Kant's lines of argumentation—a similarity nicely captured in Leibniz's phrase '...*nisi*...*Deus existeret, nihil possibile foret*'—in a rough translation, 'unless God existed, nothing would be possible' (GP VI 440).[63] But even more important to my purposes here is that Kant's commitment to his view of existence as a position rather than a predicate seems to derive mainly from drawing the consequences from the logical view of possibility based on the principle of contradiction. In connecting this to Leibniz's twofold view of existence presented in the previous section, we can say that Kant rejects the Descartes-style ontological proof in the *Beweisgrund* specifically because he disagrees with Leibniz's distinction between existence as a (real) predicate in the divine case and existence not being a (real) predicate in the created case. This, as we shall see in the next section, allows Kant to generalize the claim that existence is not a predicate.

[61] One might still wonder whether Kant thinks that possibility is grounded in the divine understanding, rather than in the divine attributes directly. The question is whether possibility in the *Beweisgrund* is grounded in the divine understanding (similar to Leibniz) or merely in the divine attributes (without any essential reference to the divine understanding). This has been the topic of recent controversy between Abaci, Yong, and Chignell in *Kantian Review* 19/1 (2014). While my intuition leans here to the Leibnizian side, I do not have a firm commitment on this question. I thank Reed Winegar for drawing my attention to this debate.

[62] As I have it in the first chapter of my *Possibility, Agency, and Individuality in Leibniz's Metaphysics*, this line of reasoning can be identified with Leibniz's proof of God's existence from the eternal truths (at least in the way it is presented in the *Monadologie* 43, *Theodicy* sec. 20, *New Essays* 4.9.14; GP V 429).

[63] GP VI 440. Leibniz's dictum is also echoed in his *Theodicy* §184: 'Sans Dieu, non seulement il n'y auroit rien d'existant mais, il n'y auroit rien de possible.' 'Without God, not only would there be nothing existing but nothing would be possible either.'

## 3.4 Kant's Critical Period: Existence and Real Possibility

In the *Critique of Pure Reason* Kant amends his previous reasoning by drawing an explicit distinction between real and logical predicates, now stressing that existence is not a *real* predicate.[64] There is also more emphasis on distinction between logical possibility and real possibility. In his *Lectures on Metaphysics* the main feature of real possibility is spelled out clearly, as follows:

> Logical possibility, actuality, and necessity are cognized according to the principle of contradiction... Real possibility is the agreement with the conditions of a possible experience.[65]

Kant's use of real possibility is not new. But, due to Kant's critical turn, it acquires a new sense. As Kant states in the passage above, real possibility does not depend on mere concepts alone, defined by the principle of contradiction (logical possibility), but also on the notion of possible experience, now constrained not only by pure logic but also by transcendental logic, that is, by appeal to our cognitive faculties, notably the forms of our sensible intuition.[66] Accordingly, whatever we can judge as existing has to fall within the realm of (real) possible experience, that is, whatever can supply the material for our considering possibilities, which can be found to be either contradictory or consistent.

As just noted, the most visible change in Kant's reasoning from 1763 to 1781 (and more broadly from the pre-critical to the critical writings) is that the dictum 'existence is not a predicate' is re-written as 'existence is not a *real* predicate'. How is this distinction related to Kant's distinction between logical and real possibility? The short answer, I think, is this: real predicates are predicates that we can ascribe to real possibilities on the basis of possible experience, that is, what we *can* (in the transcendental sense) experience rather than merely conceive the possibility.[67]

[64] This point is underestimated by Fisher and Watkins ('Kant on the Material Ground of Possibility: From *The Only Possible Argument* to the *Critique of Pure Reason*', *The Review of Metaphysics* 52/2 (1998): 369–95), who write that Kant's 'reason for rejecting the ontological argument in particular is identical to the one presented in the first Critique' (p. 370) and who suppose that the distinction is already at work in 1763. While they are right that it is anticipated in 1763 (2:72), it is clearly underdeveloped and hardly plays the same role in Kant's argument as it does in the critical work.

[65] Metaphysik Mrongovius in Immanuel Kant, *Lectures on Metaphysics,* Cambridge Edition, trans. and eds. Karl Ameriks and Steve Naragon (Cambridge: Cambridge University Press, 1997), 28:557.

[66] This point is further explained and argued for in Jessica Leech's contribution to this volume. Leech contends that we can understand real possibility in terms of conditions of cognition rather than conditions for thought alone.

[67] 'Logical possibility is possibility of a concept, and the principle of contradiction <*principium contradictionis*> is its principle criterion. Real possibility is different from this, here the principle of contradiction does not suffice. What is really possible is also logically possible but [it is] not [the case that] what is logically possible is also really possible. (The impossible is twofold: (I) when either the concept itself is nothing, e.g., four cornered circle, (II) or where no possible object corresponds, e.g., fairy tales.) Logical possibility is that wherein there is no contradiction. Metaphysical possibility is where the matter in and for itself is possible without relation to my thoughts. How am I to judge a matter, what it is in and for itself, without reference to experience?' (Metaphysik Mrongovius in Immanuel Kant, *Lectures on Metaphysics,* p. 166; 29: 811–12).

In Kant's *Lectures on Metaphysics,* section entitled On Existence (28: 554), we find the following passage:

> I cognize logical possibility through the principle of contradiction. Everything that exists is, to be sure, thoroughly determined; but with existence the thing is posited with all its predicates, and thus thoroughly determined. Existence, however, is not a concept of thoroughgoing determination; for I cannot cognize this, and omniscience is required for it.[68] Existence thus must depend not on the concept of thoroughgoing determination but rather the reverse. If something is only thought, then it is *possible.* If something is thought because it is already given, then it is actual. And if it is given because it is thought, then it is *necessary.* Through existence I think nothing more of the thing than through its possibility, but only the manner of the positing is different, namely *the relation to me.* Existence thus gives no further predicate to the thing. One says in the schools: existence is the complement <*complementum*> of possibility. But it is that which added only in my thoughts and not to the thing. The true explanation of existence is: existence is absolute positing <*existenia est positio absoluta*>. It thus can be no complement <*complementum*>, no predicate of a thing, but rather the positing of the thing with all its predicates. (28: 554; CE p. 320, italics added)[69]

In his *Lectures on the Philosophical Doctrine of Religion* Kant writes:

> We have no concept of real possibility except through existence, and in the case of every possibility which we think *realiter* we always presuppose some existence; if not the actuality of the thing itself, then at least an actuality in general which contains the data for everything possible.[70]

Existence does not add to the description of a thing but only expresses a relation to me. It is in this sense that 'Existence' is considered to be a logical but not real predicate.[71] This change is strongly related to Kant's qualification of existence as a logical (but not real) predicate. A real predicate has to add a certain (positive) reality or content to the nature of the thing. A logical predicate, however, describes our judgement about the modal status of a concept. Whereas logical possibility is defined by mere consistency among terms in a concept, real possibility has to do with an object (denoted by the concept) and thus given in possible experience.[72] This is why, as Kant says, 'one can infer to possibility from existence, but not the reverse, to existence from possibility' (*Lectures on Metaphysics* 28: 555; CE p. 320).

[68] In Leibniz, of course, this is precisely what is being presupposed with the notion of an omniscient mind (God) who is actually thinking all possibilities.

[69] See also his discussion of logical and real essence in 28:553.

[70] Cited from Adams, 'God, Possibility, and Kant', p. 428.

[71] See Abaci, 'Kant's Theses on Existence', *British Journal for the History of Philosophy* 16/3 (2008): 559–93, note 18 for a convincing account of the transition from 1763 to Kant's critical conception of modality.

[72] As Jessica Leech puts this in section 2 of her article in this volume: 'Logical possibility is to be understood in terms of the conditions for being able to think something (laws of thinking); real possibility is to be understood in terms of the conditions for being able to cognize something (laws of cognition).' The possibility of being able to cognize something is clearly constrained in the Critique by the subjective conditions of experience.

Indeed, it is precisely the move from (the possibility of) concepts to the existence of objects that is the crux of Kant's critique of the ontological argument. In the following footnote Kant makes this point clearly:

> The concept is always possible if it does not contradict itself. That is the logical mark of possibility... Yet it can nonetheless be an empty concept, if the objective reality of the synthesis through which the concept is generated has not been established in particular; but... this always rests on principles of possible experience and not on the principles of analysis (on the principle of contradiction). This is a warning not to infer immediately from the possibility of the concept (logical possibility) to the possibility of the thing (real possibility)
>
> (*Critique*, A596/B624; GW p. 566).

While the possibility of concepts depends on (internal relations between) concepts alone, the possibility of objects depends on experience (and its constraints) as well. It is for this reason that Kant says: 'You have already committed a contradiction when you have brought the concept of its existence, under whatever distinguished name, into the concept of a thing which you would think merely in terms of its possibility' (B625/A597; GW p. 566). Since judgements of modality can only pertain to the thing with all its predicates (its complete concept), existence cannot be introduced into it. If a concept of a certain thing would posses another predicate, be it 'existence' or any other, it would be a concept of a different thing (A600/B628; GW pp. 567–8).

While Kant has adapted a version of the Leibnizian notion of a complete concept of the individual, seen as a pure possibility,[73] the implicit target here (as Kant makes clear in the final paragraph to this section of the *Critique*) is precisely Leibniz's attempt to show that existence follows from the very possibility of a concept that entails all perfections (the *ens perfectissimum*). Kant points out that this very attempt is contradictory because the notion of existence does not belong to the logical possibility or the internal consistency of a concept, which must be already thoroughly determined in order that *it* could be considered as existing or not (A600/B628).

An existence claim does not express a (determination) relation between a subject and one of its real predicates; rather, it expresses a relation between a subject that already includes all its predicates to a mind considering its position among the modal categories (possible, actual, or necessary). Since judgements of existence do not add any determination to the content of a concept but rather express its status, i.e., whether it is actual or possible, they belong to the category of modality, which does not include real but only logical predicates.

As Kant notes explicitly:

> Possibility, actuality, necessity are not concepts of things in themselves; rather possibility already presupposes the thing with all its predicates, and the comparison of the thing with

[73] It is worth noting, however, that Kant's complete concept is not the same as Leibniz's complete individual concept. Kant denies the possibility of a complete individual concept of any empirical thing, because spatial and temporal properties are not conceptual.

the laws of thinking, whether it can be thought or not. (Actuality is that to which an object of experience corresponds; necessity is actuality that follows from possibility.)

(*Lectures on Metaphysics* 29:822, Metaphysik Mrongovius, pp. 175–6)

As Kant notes explicitly, modal terms are 'not real predicates or determinations but rather logical ones, e.g., God is possible.' (*ibid.* p. 176; and see also 28:554).[74] Thus, the point he stresses here is that existence is a modal/logical predicate, not a real one. Real predicates express realities or positive attributes of possible individuals. Interestingly, they are reminiscent of what Leibniz calls positive attributes or perfections, which are presupposed as the basis of his system of possibility. Kant's notion of the *ens realissimum*—the notion of an individual consisting of all positive realities—is likewise strongly related to Leibniz's notion of the *ens perfectissimum*, an individual being that entails all positive perfections and that serves as the ground of all possibilities.[75]

However, in opposition to Leibniz, Kant's point in the *Critique* is precisely that this very notion, the most perfect or supreme being, while it appears to make a claim for reality, is rather a mere idea. It is thus not a being that has to be presupposed as the ground of possibility—the individual containing all possible (positive) predicates but only an idea of such a being or such notion of all reality (*omnitudo realitatis*) as a demand of reason (A575–6/B603–4):

The concept of a highest being is a very useful idea in many respects; but just because it is merely an idea, it is entirely incapable all by itself of extending our cognition in regard to what exists. It is not even able to do so much as to instruct us in regard to the possibility of anything more. The analytic mark of possibility, which consists in the fact that mere positings (realities) do not generate a contradiction, of course, cannot be denied of this concept; since, however, the connection of all real properties in a thing is a synthesis about whose possibility we cannot judge a priori because the realities are not given to us specifically—and even if this were to happen no judgment at all could take place because the mark of possibility of synthetic cognitions always has to be sought only in experience, to which, however, the object of an idea can never belong—the famous Leibniz was far from having achieved what he flattered

[74] See *Critique*, A234: 'The principles of modality are not, however, objective-synthetic, since the predicates of possibility, actuality, and necessity do not in the least augment the concept of which they are asserted in such a way as to add something to the representation of the object. But since they are nevertheless always synthetic, they are so only subjectively, i.e., they add to the concept of a thing (the real), about which they do not otherwise say anything, the cognitive power whence it arises and has its seat, so that, if it is merely connected in the understanding with the formal conditions of experience, its object is called possible; if it is in connection with perception (sensation, as the matter of the senses), and through this determined by means of the understanding, then the object is actual; and if it is determined through the connection of perceptions in accordance with concepts, then the object is called necessary.' (GW pp. 332–3).

[75] 'The derivation of all other possibility from this original being... cannot be regarded as a **limitation** of its highest reality and as a **division**, as it were, of it; for then the original being would be regarded as a mere aggregate of derivative beings... Rather, the highest reality would ground the possibility of all things as a **ground** and not as a **sum total**; and the manifoldness of the former rests not on the limitation of the original being itself, but on its complete consequences.' (A579/B607; GW p. 557).

himself he had done, namely, gaining insight a priori into the possibility of such a sublime ideal being. (B630/A602; GW pp. 568–9)

While the logical possibility of the idea of a supreme being cannot be denied, its real (synthetic) possibility could never be ascertained a priori since this would require an appeal to experience. Real possibility requires possible cognition rather than mere thought. Existence claims cannot be analytic but must refer to some (possible) intuitive content. Hence, for Kant, Leibniz's project of proving a priori the *existence* of the *Ens Perfectissimum* from its logical possibility is doomed. The adequate status of this concept in Kant's critical philosophy is that of an ideal of reason—an idea that must still be presupposed at the background of his theory of possibility (A578/B606).[76]

## 3.5 Conclusion

The picture at the background of this complex story can be now presented along the following lines: the acceptance of Kant's point that existence is not a predicate requires a new picture of the relations between essence and existence. Roughly stated, in such a picture the notion of essence is understood in terms of pure logical possibility or, more precisely, in terms of the consistency relations between concepts—relations that make no reference to existence, time or place. While the seeds of such a conception were around for a long time (at least since Scotus and probably earlier), an explicit and influential identification of the essence of an individual with a complete concept that specifies every truth about it, which also allows it to be considered as a possible candidate for actualization, was developed and systematically employed by Leibniz.

Leibniz carried out the programme of divorcing the essence or the possibility of an individual from its existence a long way but he stopped at God, in whose concept essence and existence are seen to be inseparable (or as conceptually related). This is precisely what Leibniz's a priori proof for the existence of God purports to show. More precisely, this is what Leibniz's proof that such a being is *possible*, purports to show. It is arguable that Leibniz's reasons to stop at God are in the end his theological commitments. As I pointed out, though, Leibniz had some deep philosophical

[76] The role of this idea in Kant's practical philosophy is of great consequence but this is a topic that goes beyond the scope of the present article. 'One cannot directly prove the *existence* of any thing a priori, neither by an analytic nor a synthetic principle of judgement. To assume it, however, as a hypothetical thing for the sake of possible appearances, is to feign, not to demonstrate, *cogitabile non dabile*. The concept of God is, however, the concept of a being that can *obligate all moral beings* without itself [being] obligated, and, hence, has a rightful power over them all. To wish to prove the *existence* of such a being *directly*, however, contains a contradiction, for *a posse ad esse non valet consequentia*. Thus only an *indirect* proof remains, inasmuch as it is assumed that something else be possible, namely, that the knowledge of our duties *as (tanquam)* divine commands is certified and authorized—not in theoretical but in a pure practical respect—as a principle of practical reason, in which there is valid inference from *ought to can*' (22:121 *Opus Postumum*, CE, edited by Eckart Förster, translated by Eckart Förster and Michael Rosen, 1993, p. 203).

reasons to maintain this position as well. It turns out that some of these reasons—especially that the possible presupposes something actual—are endorsed by Kant in his pre-critical essay of 1763, where he already states that existence is not a predicate (and uses the notion of the *ens realissimum*).[77] Likewise, some of these Leibnizian considerations are explicit in Kant's (1781) critique of any attempt to prove existence claims—first and foremost, God's existence—from concepts alone. In terms of Kant's 1763 essay, internal possibility itself presupposes something actual, which, according to Kant, is the only foundation for a possible proof of God's existence. In terms of the first *Critique*, any claim for existence requires synthetic judgements and, for this reason, existence claims cannot be known a priori. Thus, the general notion of logical possibility as self-consistency among concepts seems to be one of Kant's fundamental commitments. It is this commitment, I suggest, that is behind Kant's generalization of the notion of existence as indicating the modal position—possible, actual, or necessary—of a concept. Kant's commitment to Leibniz's logical view of possibility is an essential piece of the background of his view that existence is not a real predicate, so that it cannot add any content to a concept. As we have seen, the same point also applies to the other modal judgements of actuality and necessity. In this respect, the Leibnizian view of possibility, as accepted and developed by Kant, helps to explain his position, already evident in 1763, that existence is not a predicate. This is the main point I have tried to establish in the first three sections.

In the *Critique*, however, Kant makes another radical move: he argues explicitly that experience is necessary in order to satisfy the material condition of possibility. While experience has already been referred to in the pre-critical period, the notion of experience is now significantly modified. The notion of real possibility is referred to that of experience, now understood as necessarily constrained by the subjective conditions of human cognition. In thus referring the material of thought to *human* sensibility, Kant leaves behind him another substantial Leibnizian commitment, namely that possibilities are conceived in God's understanding and thus presuppose it as their ultimate ground. Strictly speaking, this pertains to the notion of real possibility rather than to mere conceivability of an understanding unbounded by intuition—something we can merely conceive as a possibility, and thus this notion might remain conceivable for things in themselves, which are supposed to be merely thinkable and inaccessible to any sensibility. The distinction between logical and real possibility might also account for Kant's notion of things in themselves. It explains why things in themselves can be thought but not cognized, so that we can conceive their mere possibility but never make any claims about such things beyond their possibility. Another way of putting this point is that, in the critical period, modality is relativized to a specific understanding: while real possibility pertains to human

[77] This notion deserves more attention. For a brief discussion of Kant's notion see Wood, *Kant's Rational Theology*, pp. 55–9.

understanding, pure logical possibility, independent of any sensibility may still be identified with God (purely conceptual understanding).[78]

As I noted earlier, Leibniz's theory of possibility presupposes God's mind and his attributes as its necessary actual grounds.[79] According to Leibniz, if the notion of God is possible, it follows that God also exists. According to Kant, of 1763, if something is possible, then there exists a necessary being that constitutes the ground for possibility in general. My main point here has been to argue that, for both Leibniz and Kant, the notion of possibility as consistent thinkability constitutes the most fundamental commitment. But this point must be finally qualified: while, for Leibniz, thinkability refers primarily to the divine understanding, which is omniscient and absolute, for the critical Kant, modality judgements refer primarily to human thinkability, which is constrained by its subjective conditions and limitations. But these limitations, due to our sensibility, do not pertain to the form of thought (and thus allow the possibility of purely formal objects of thought and thus for the possibility of things in themselves); they restrict the possible material of human thought and judgement. While human reason remains for Kant universal and purely formal, the content or the material of its thoughts—what must be given to us through intuition—requires reference to human intuition—that is, the material obtained through our peculiar subjective cognitive faculties. Otherwise, formal objects that necessarily lie beyond possible experience fall into the realm of problematic pure concepts, thinkable but not sensible, things in themselves.

In thus referring real possibility to the human rather than to the divine mind Kant moves beyond the Leibnizian framework of conceiving possibilities. Likewise, it is for this very reason (the necessary supposition of God's mind as the ground of thinkability and possibility) that Leibniz would not, and arguably could not generalize the concept of existence as a mere actualization of a complete concept. As Kant moves away from conceiving of real possibilities in terms of divine thinkability, he is no longer obliged to make any exceptions for the notion of existence as well. In generalizing the scope of existence as a modal position—actual rather than merely possible, and thus not a predicate or anything that belongs to the realm of concepts, Kant completes what Leibniz has begun. In this way, 'the Critique of Pure Reason might well be the true apology for Leibniz'.[80]

[78] Reed Winegar notes that Kant is very sympathetic to this kind of principle. He states a version of it in the Inaugural Dissertation to the effect that that things-in-themselves need to be cognizable by some kind of understanding in order to be regarded as possible and, thus, construes things-in-themselves (both in the Inaugural Dissertation and elsewhere) as the objects of a possible intuitive understanding. In the 3rd Critique Kant emphasizes that the possibility of a unity of mechanism and teleology requires that this unity be cognizable by some kind of understanding, which again is proposed to be an intuitive understanding. In short, Kant seems extremely suspicious of saying that some thing could be possible even though there is no type of understanding that could cognize it.

[79] See chapters 1 and 2 of my *Possibility, Agency, and Individuality in Leibniz's Metaphysics.*

[80] As Kant says in the end of his polemical essay 'On a discovery whereby any new critique of pure reason is to be made superfluous by an older one' (1790). This might account for another puzzle viz., why

Yet there is an ironic addendum to this story—an ironic closing of a circle, so to speak. I started the paper by arguing that Leibniz replaced the temporal notion of possibility with a logical one. But, if, as we have just seen, the subjective conditions of human sensibility, space and time, play a constitutive role in providing the material for our modal judgements concerning real possibilities, then it would seem that Kant's theory of modality in his critical period goes back full circle to a temporal notion of possibility. It is not exactly a full circle, though. After all, Kant's notion of temporality, seen as a form of our intuition constitutive of anything we can experience, is radically different from the Aristotelian notion of temporality rejected by the champions of logical possibility—and especially by Leibniz. This topic, however fascinating, must wait for another occasion.[81]

## A note on the method of citation

Citations from Kant's *Critique of Pure Reason* are given by reference to the pagination of the first (A) edition (1781) and/or the second (B) edition (1787). Citations from all Kant's other works are given by the volume and page number, separated by a colon, in the standard edition of Kant's work, the Academy edition (*Gesammelte Schriften ed. Königlich Preussische Akademie der Wissenschaften* (Berlin: de Gruyter, 1902–), Volumes 1–29). Unless otherwise indicated, all translations are from the *Cambridge Edition of the Works of Immanuel Kant,* General Editors, Paul Guyer and Allen Wood, (1992–), abbreviated as CE. Citations from Leibniz's works are given by reference to the standard Academy edition (A) and by reference to the following standard abbreviations and translations.

## Abbreviations

A Leibniz, G. W. *Sämtliche Schriften und Briefe,* Darmstadt/Leipzig/Berlin Edition of the Berlin Academy, 1923–. Cited by series volume and page.

AG Leibniz, G. W. *Philosophical Essays*, edited and translated by Roger Ariew and Daniel Garber (Indianapolis: Hackett, 1989).

CP Sleigh, R. C., and Sleigh, J. R., (ed. and trans.), *Confessio philosophi, Papers Concerning the Problem of Evil, 1671–1678* (New Haven: Yale University Press, 2005).

GP *Die Philosophichen Schriften von Leibniz,* edited by Carl I. Gerhardt, 7 vols. (Berlin: Weidmann, 1875–90); reprinted (Hildesheim: Olms, 1978).

Kant has not returned to his argument of 1763 for the existence of God as the ground of possibility. Once the notion of possibility is relativized to an understanding (human or divine or whatever) supposing God's understanding as the sole and necessary grounds for possibility no longer makes sense.

[81] I thank Reed Winegar once again for making me see this point.

L Leibniz, G. W., *Philosophical Papers and Letters*, translated and edited by L. Loemker (Dordrecht: Reidel, 1969).

LR Leibniz, G. W., *Discours de métaphysique et correspondance avec Arnauld*, edited by George Le Roy (Paris: Vrin, 1970).

DSR Leibniz, G. W., *De Summa Rerum: Metaphysical Papers 1675–1676*, translated and edited by G. H. R. Parkinson (New Haven and London: Yale University Press, 1992).

## Bibliography

Abaci, U., 'Kant's Only Possible Argument and Chignell's Real Harmony', *Kantian Review*, 19/1 (2014): 1–25.

Adams, R. M., *Leibniz: Determinist, Theist, Idealist* (New York: Oxford University Press, 1994).

Adams, R. M., 'God, Possibility, and Kant', *Faith and Philosophy*, 17/4 (2000): 425–40.

Antognazza, M. R., 'Arguments for the Existence of God: The Continental European Debate', in *The Cambridge History of Eighteenth-Century Philosophy*, edited by K. Haakonssen (Cambridge: Cambridge University Press, 2006), 731–48.

Chignell, A., 'Kant and the "Monstrous" Ground of Possibility: A Reply to Abaci and Yong', *Kantian Review*, 19/1 (2014): 53–69.

Fichant, M., 'L'origine de la négation', in Fichant, M., *Science et métaphysique dans Descartes et Leibniz* (Paris: Presses Universitaires de France, 1998), 85–119.

Fisher, M., and Watkins, E., 'Kant on the Material Ground of Possibility: From *The Only Possible Argument* to the *Critique of Pure Reason*', *The Review of Metaphysics*, 52/2 (1998): 369–95.

Hintikka, S., 'Leibniz on Plenitude, Relations, and the "Reign of Law"', in *Reforging the Great Chain of Being*, edited by S. Knuuttila (Dordrecht: Springer, 1981), 259–86.

Ishiguro, H., *Leibniz's Philosophy of Logic and Language*, 2nd edition (Cambridge: Cambridge University Press, 1990).

Jolley, N., (ed.), *The Cambridge Companion to Leibniz* (New York, 1995).

Kannisto, T., 'Positio contra complementum possibilitatis—Kant and Baumgarten on Existence', *Kant-Studien*, 107/2 (2016): 291–313.

Knuuttila, S., 'Modal Logic', in *The Cambridge History of Later Medieval Philosophy*, edited by N. Kretzmann, A. Kenny, and J. Pinborg (Cambridge: Cambridge University Press, 1982), 342–57.

Laerke, M., *Leibniz lecteur de Spinoza. La genése d'une opposition complexe* (Paris: Honoré Champion, 2008).

Mates, B., *The Philosophy of Leibniz: Metaphysics and Language* (New York: Oxford University Press, 1986).

Mondadori, F., 'Modalities, Representations, and Exemplars: The "Region of Ideas"', in *Mathesis rationis*, edited by A. Heinekamp, W. Lenzen, and M. Schneider (Münster: Dutz, 1990), 169–88.

Mugnai, M., 'Leibniz's Nominalism and the Reality of Ideas in the Mind of God', in *Mathesis rationis*, edited by A. Heinekamp, W. Lenzen, and M. Schneider (Münster: Dutz, 1990), 153–67.

Mugnai, M., 'Leibniz's Theory of Relations', *Studia Leibnitiana Supplementa*, 28 (Stuttgart: Franz Steiner, 1992).

Oppy, G., *Ontological Arguments and Belief in God* (Cambridge: Cambridge University Press, 1995).

Russell, B., *A Critical Exposition of the Philosophy of Leibniz*, 2nd edition (London: Allen and Unwin, 1937).

Rutherford, D., 'Language and Philosophy in Leibniz', in *The Cambridge Companion to Leibniz*, edited by N. Jolley (New York: Cambridge University Press, 1995).

Vilkko, R., and Hintikka, J., 'Existence and Predication from Aristotle to Frege', *Philosophy and Phenomenological Research*, LXXIII/2 (2006): 359–77.

Vilmer, J. J.-B., 'L'existence leibnizienne', *Archives de Philosophie*, 70 (2007): 249–72.

Wood, A., *Kant's Rational Theology* (Ithaca: Cornell University Press, 1978).

Wood, A., 'Rational Theology, Moral Faith, and Religion', in *The Cambridge Companion to Kant*, edited by P. Guyer (Cambridge: Cambridge University Press, 1992).

Yong, P., 'God, Totality and Possibility in Kant's Only Possible Argument,' *Kantian Review*, 19/1 (2014): 27–51.

# 4

# Kant's Material Condition of Real Possibility

*Jessica Leech*

In recent years, there has been a surge of interest in Kant's views on modality.[1] Why? Whilst it is true that this is an underexplored area of Kant scholarship there is also significant independent philosophical interest in exploring an approach to understanding modality which avoids the excesses of realism and anti-realism. Baldwin (2002) captures this spirit nicely. Having discussed reasons for dissatisfaction with a Lewisian possible worlds approach, he moves on to consider a Kantian approach as an attractive middle way between an Aristotelian realism and a Humean anti-realism:

> Of the alternatives to full-blooded realism, the Kantian position appears, to me at least, *prima facie* the most attractive. For it offers the prospect of an account which does not treat modality as a primitive feature of reality, in the way that an appeal to Aristotelian essences appears to, while equally avoiding the subjectivism of Humean projectivism.[2]

I share this general perspective on the problem of understanding modality. On the one hand, I think there are serious barriers to making proper sense of a notion of the essence of an object that can play any helpful role in our understanding of modality.[3] On the other, I grant that there do seem to be modal matters of fact that go beyond a

[1] See, for example, U. Abaci, 'The Coextensiveness Thesis and Kant's Modal Agnosticism in the "Postulates"', *European Journal of Philosophy* 24:1 (March 2016): 129–58; T. Baldwin, 'The Inaugural Address: Kantian Modality', *Proceedings of the Aristotelian Society*, Supplementary Volumes 76 (2002): 1–24; A. Chignell, 'Kant, Modality, and the Most Real Being', *Archiv für Geschichte der Philosophie* 91 (2009): 157–92, 'Kant, Real Possibility, and the Threat of Spinoza', *Mind* 121(2012): 635–75, and 'Kant and the "Monstrous" Ground of Possibility', *Kantian Review* 19:1 (2014): 53–69; M. Fisher, and E. Watkins, 'Kant on the Material Ground of Possibility: From *The Only Possible Argument* to *The Critique of Pure Reason*', *Review of Metaphysics* 52 (1998): 369–95; J. Leech, 'Kant's Modalities of Judgment', *European Journal of Philosophy* 20 (2012): 260–84, and 'Making Modal Distinctions: Kant on the Possible, the Actual, and the Intuitive Understanding', *Kantian Review* 19:3 (2014): 339–65; N. F. Stang, 'Kant's Possibility Proof', *History of Philosophy Quarterly* 27 (2010): 275–99, 'Did Kant Conflate the Necessary and the A Priori?', *Nous* 45 (2011): 443–71, and *Kant's Modal Metaphysics*, Oxford University Press (2016).

[2] Baldwin, 'The Inaugural Address: Kantian Modality', p. 9.

[3] I develop this argument in other work, currently in progress.

mere projection of attitudes onto a world which is not at all modal in itself. A Kantian approach can make appeal to aspects of the experiencing subject to explain modality, and hence avoid a thoroughgoing realism, whilst at the same time maintaining that these aspects feed into what the world is really like, hence avoiding an overly subjective view.

Taking a generally Kantian approach is all very well, but that approach then needs to be worked out in detail. One way to embark upon such a project is to examine first what Kant himself wrote about modality, in the hope that the details of his views may point us towards a developed Kantian account of modality. This is a project that Baldwin sets to one side—'in exploring a "Kantian" position . . . I am not concerned to follow the details of Kant's actual discussion of modality as a category'[4]—but I think we can take this as an invitation to take on that task, rather than as a recommendation to ignore it.

The aim of this chapter is thus to continue this project, and to explore some of the details of Kant's thinking about modality, in the hope that we can learn something about modality (and not just about Kant). I want to focus in particular on how to incorporate a particularly puzzling passage into our understanding of Kant's views. We can gain a fairly clear account of real possibility in terms of formal conditions of experience from the section of the *Critique of Pure Reason* which is devoted to the categories of modality (the *Postulates of Empirical Thinking in General*), but much later in the *Critique* we suddenly encounter a new claim about the *material* of all possibility of empirical objects. At this point, it's not at all clear what Kant means, let alone how it fits in with his earlier discussion of modality. The plan of this chapter is thus as follows. First, I'll sketch my understanding of the key background view of real possibility, its connection to Kant's account of cognition, and how it contrasts with logical possibility. I'll then move on to sketch the argument of the section which leads up to the puzzling claim. We'll see that Kant argues that rational theology is based on a misunderstanding of an idea of reason, and also a mistake in taking this idea of reason for something else—the ground of the possibility of all empirical objects. I will argue that these issues all turn on Kant's account of how we can individuate objects. The additional puzzling claim about real possibility will turn out to arise naturally from Kant's overall connection between real possibility and cognition, just as in the case of his account of modality earlier in the *Postulates*.

## 4.1 Kant on Cognition

In the *Critique of Pure Reason*, Kant presents an account of those of our cognitive capacities which contribute to representation and knowledge. *First*, he distinguishes between two basic capacities we have for representing things: *sensibility*, a passive or

[4] Baldwin, 'The Inaugural Address: Kantian Modality', p. 9.

receptive capacity for intuition; and *understanding*, an active or spontaneous capacity for conceptual thought. Intuitions are immediate, singular presentations of things; concepts are mediate, general presentations of things.[5] In other words, Kant assumes that we are both sensible creatures—we rely on our senses to give us information about the outside world—and conceptual creatures—we can think about things as well as sense them. *Second*, we need both of these capacities in order to have experiences of the world. Concepts and intuitions cooperate to form *cognitions*: objective representations of the empirical world. These two kinds of representations need each other to produce a representation which is genuinely about an object. Kant sums this up in his famous line:

> Thoughts without content are empty, intuitions without concepts are blind. (A51/B75)[6]

A concept needs to be given something to apply to (something to be about). An intuition needs to be conceptualized in order to form a representation of something *as* something.

According to Kant, certain of our representations are prerequisites of the very possibility of having experience at all. So they can't be *derived* from experience. Rather, they provide the very conditions under which we can have experience in the first place. The pure representations of sensibility are the pure forms of intuition: Kant argues that the forms of space and time are preconditions of intuitions of things. The pure representations of the understanding are pure concepts or 'categories': Kant argues that concepts such as *magnitude*, *substance*, and *causation* combine to provide 'the concept of an object in general',[7] a formal framework in which our experience is organized into one of physical objects with qualitative properties, causally interacting, subject to laws of physics, and so on. A thought, then, must conform to these conditions on possible experience—forms of intuition and principles arising from the categories—if it is to properly relate to the world of experience, and be a cognition. Experience, for Kant, is cognition based on empirical intuition, i.e. intuitions arising from sensation, from input given through the senses. Synthetic a priori knowledge is also possible by taking into account only the pure conditions of cognition. For example, Kant took mathematical knowledge to be synthetic a priori, explained in terms of relying only on pure intuition (not empirical intuition).

One important division of labour between concepts and intuitions is that, for Kant, it is intuitions that individuate things, i.e. intuitions are able to represent individuals, whereas concepts are by nature general representations. Of course, it can happen that sometimes we succeed in picking out individuals using concepts, and Kant allows for this.[8] For

[5] See (A19/B33), (A320/B337).

[6] A/B numbers indicate citations from the *Critique of Pure Reason*, translated and edited by Paul Guyer and Allen W. Wood (New York: Cambridge University Press, 1998).

[7] (A51/B75, B128). See also (A80/B106) for the full table of categories.

[8] 'It is a mere tautology to speak of general [*allgemeinen*] or common concepts—a mistake that is grounded in an incorrect division of concepts into *general* [*allgemeine*], *particular* and *singular*. Concepts

example, the concept *red kitchen utensil* will succeed in picking out a unique individual only if there is only one red kitchen utensil, and under such circumstances such a concept could be used to refer to an individual. But there is no *guarantee* in a conceptual representation that it will pick out an individual. For Kant, that is the work of intuition.

This is borne out by Kant's rejection of the identity of indiscernibles (in the *Amphiboly* section of the *Critique*). Suppose that our only means of representation is through concepts. We can then build concepts that are maximally specific, i.e., for any constituent concept, our maximal concept either includes it or its negation. Now, if we use such concepts to pick out objects *a* and *b*, and if those maximally specific concepts are the same, i.e. all and only the same concepts apply to *a* and *b*, then there is nothing to tell between *a* and *b*, and one may conclude that *a* = *b*. However, Kant claims that we also need sensibility—we represent objects through a combination of conceptual *and sensible* means. Hence, two things could fall under all and only the same concepts, having the same maximally specific concept, and yet be distinguishable through being presented in distinct intuitions. Given that the form of intuition is space and time, this amounts to the simple claim that otherwise indistinguishable things are distinct insofar as they have different spatiotemporal locations.

> Thus, in the case of two drops of water one can completely abstract from all inner difference (of quality and quantity), and it is enough that they be intuited in different places at the same time in order for them to be held to be numerically different. (A264–5/B319–20)

We can think about all sorts of things, more or less specific, using our concepts. But if we want to home in on a particular individual, then we need intuition. Moreover, not only can a concept alone not guarantee that only one object falls under it, a concept cannot guarantee that anything at all falls under it. Again, for that the input of intuition is required. As I put it above, for Kant, a concept needs to be given something to apply to.[9]

One might object that there are certain concepts which could only have one object falling under them, and thus that such concepts guarantee individual reference, e.g. (1) 'the one and only brother of Christopher Columbus', (2) 'the smallest prime number', (3) 'the tallest tree', (4) 'the only contingent object in existence'.

Example (1) would indeed succeed in referring to an actual object only if there were one and only one brother of Columbus. However, it would only succeed in picking out a particular individual insofar as it piggy-backs upon an already given object—Columbus actually existed. So this isn't a case of purely conceptual individuation. Moreover, the concept alone doesn't guarantee that there is anything falling

themselves cannot be so divided, but only *their use*.' (Kant, I., *Lectures on Logic*, translated and edited by J. Michael Young (New York: Cambridge University Press, 1992), 589; 9:91, translation amended).

[9] The alternative would be that the existence of an individual falling under a concept could be guaranteed by the concept alone. This would require something more like an intuitive intellect. See Leech, 'Making Modal Distinctions' for a more detailed discussion of the differences between cognitive capacities which do, and do not, require sensible intuition.

under it. For that intuition is still required. Perhaps we could still assess such a concept for real possibility, but again, it seems that we are only considering an individual possibility insofar as it is connected to something actual, and hence already individuated through intuition.

Example (2) appears to be more tricky: one might argue that there is only one object falling under it by conceptual necessity. The purist Kantian response would be to point out that, for Kant, such mathematical examples are to be understood as having their source in the pure form of intuition, and so insofar as such an example picks out an individual, that is only because it depends upon the form of intuition, and not on anything conceptual. But one may be wary of a commitment to Kant's views on mathematics.

In any case, examples such as (3) and (4) still pose problems that are not as easily dismissed as (1). It seems to me that, with contemporary advances in our understanding of logic and language, we can't defend Kant's claim that a concept can never be individual in the sense that, if satisfied, it can only be satisfied by one thing. However, I don't think giving up this claim does much to diminish the key points. First, we can retain the claim that cognition of an *actual* individual requires intuition, even if we can perhaps assess individual concepts for *real possibility*. So, for example, it may be that it is really possible for there to be a tallest tree, but there might not actually be one, and we could only cognize the actual tallest tree through empirical intuition of it (or something causally related to it).[10] Second, it remains the case that the vast majority of our concepts are not individuating. Whilst there may be a few examples as listed above, most concepts, however specific, won't guarantee individuation. Hence we can still see the roles of intuition and spatiotemporal location as vital elements of Kant's account of individuation, even if in rare cases concepts are more powerful than Kant allowed for.

## 4.2 Kant on Logical and Real Possibility

In Kant's view, something is *logically possible* just when the concept of it is non-contradictory, and something is *really possible* just when the concept of it is non-contradictory *and* consistent with the a priori constraints on experience arising from the forms of intuition and the categories.

> *Logical* possibility, actuality, and necessity are cognized according to the principle of contradiction [...] Real possibility is the agreement with the conditions of a possible experience.[11]

[10] See A226/B273.

[11] Kant, 'Metaphysik L2' in *Lectures in Metaphysics*, edited by K. Ameriks and S. Naragon (Cambridge: Cambridge University Press, 1997), 299–354; 28:557. See also (Bxxvi & footnote) and (A244/B302 & footnote), but this statement from Kant's lectures on metaphysics is particularly clear.

For example, even though the concept of a causally-isolated part of the universe plausibly isn't a contradiction in terms, and is thus a logical possibility, if it is incompatible with conditions on possible experience,[12] then it will turn out to be really impossible.

Kant develops his account of real modality in the *Postulates of Empirical Thinking in General*. This section of the *Critique* is part of the *Analytic of Principles*, where Kant fleshes out the content of the categories. In so doing, he also provides specific arguments for why each such concept is a necessary condition on possible experience. Kant gives an account of the three modal categories—*possibility*, *actuality*, and *necessity*—in terms of relations to conditions on experience. Possibility is to be understood in terms of agreement with formal conditions on experience; actuality in terms of connection with material conditions of experience; and necessity in terms of determination in accordance with general conditions of experience.

1. Whatever agrees with the formal conditions of experience (in accordance with intuition and concepts) is **possible**.
2. That which is connected with the material conditions of experience (of sensation) is **actual**.
3. That whose connection with the actual is determined in accordance with general conditions of experience is (exists) **necessarily**. (A218/B265)

There is much of interest to be said about all three postulates, but in this paper I will focus my attention on possibility and, to a lesser extent, actuality.

Consider the postulate of possibility. The formal conditions of experience are those described above: forms of intuition (space and time) and the framework provided by the categories. The notion of agreement suggests a relation of compatibility. It is natural to understand this agreement or compatibility as logical compatibility: possible things are not in logical contradiction with formal conditions on possible experience.[13]

[12] Indeed, it is arguably incompatible with the principle of the third analogy: 'All substances, insofar as they can be perceived in space as simultaneous, are in thoroughgoing interaction.' (B256)

[13] More care is needed here. Although logical compatibility is a natural choice, this may come into conflict with other core features of Kant's view, in particular, that geometrical reasoning is synthetic, not analytic. Stang writes: 'If we analyzed formal possibility and necessity in terms of logical grounding [entailment], it would follow that geometrical axioms logically entail geometrical theorems, i.e., that all formally impossible propositions are logically incompatible with the axioms. But Kant holds that geometric reasoning is irreducibly synthetic because it relies on construction in pure intuition. Consequently the relation between forms of experience and what is formally necessary [and possible] cannot be assimilated to the relation of logical entailment; the formal necessities do not follow logically from the forms of experience.' (Stang 2016, 207). Stang takes the relevant relations to be understood in terms of real grounding, a non-logical explanation of things. Two brief points. First, although Stang offers compelling evidence for Kant's understanding of grounding from elsewhere, there is little evidence of his thinking in terms of real grounding in the Postulates chapter itself. Second, my main line of reasoning in this chapter does not turn on how we understand this relation, so I will leave further discussion for elsewhere.

What is it that is logically compatible (or logically incompatible) with formal conditions on possible experience? These modal concepts are supposed to concern the modality of *things* (A219/B267), or states of things (A227/B279–80). But objects, properties and/or events are not typically thought to be the things that stand in logical relations. Moreover, in the case of merely possible objects, there is no object to stand in the relation: the point is precisely that there is no such thing but that there *could* be. Hence, the second *relata* (the first being formal conditions on possible experience) must be a representation. Kant seems to have in mind the *concept* of the things/states, e.g. if the concept of a thing is not actually satisfied, but is nevertheless compatible with formal conditions on possible experience, then the kind of thing the concept describes is possible. Simplifying, we can say that something is really possible if *the proposition that something falls under the concept of the thing* is compatible with the formal conditions on experience.[14]

Compatibility with formal conditions on experience is thus a condition for real possibility. These formal conditions on experience are just the formal conditions for cognition. Hence we can understand real possibility in terms of conditions of cognition: conditions for a thought to genuinely be about the world (what Kant calls 'objective validity').

We can plausibly understand Kant as taking the view that logic is the study of the laws of thinking in general, in other words, Kant understood logic to be a science of the laws of thought.

> In logic we do not want to know how the understanding is and does think and how it has previously proceeded in thought, but rather how it ought to proceed in thought.[15]
>
> [T]he boundaries of logic, however, are determined quite precisely by the fact that logic is the science that exhaustively presents and strictly proves nothing but the formal rules of all thinking... (Bix)

Thinking is what the understanding does, i.e. our capacity for conceptual thought: 'The faculty for **thinking** of objects of sensible intuition... is the **understanding**' (A51/B75). Sensibility and understanding need to cooperate to generate cognitions, thoughts genuinely about objects, but Kant maintains the importance of nevertheless noting that there are two distinct capacities at work. Thinking is what the understanding does: it organizes and conceptualizes input from sensibility. The general rules for the understanding, for thinking, are the concern of *logic*. In particular, *pure general logic* concerns a priori rules for thinking, regardless of what the thinking is about. Input from empirical experience is not taken into account: this is just how the understanding functions, regardless of empirical input or constraint.[16] If Kant's view

[14] There seems to be no reason to restrict the range of propositions to those of the form 'something falls under concept C', so in developing a more general Kantian account of modality we can take the state of affairs described by any proposition to be really possible or really impossible according to whether the proposition is logically compatible or logically incompatible with conditions on possible experience.

[15] Kant, *Lectures on Logic*, 529; 9:14.

[16] See A54/B78.

of logical laws is that they have their source in the laws of thought, then we can in turn understand logical possibility, defined in terms of logical contradiction, to be a matter of compatibility with these laws of thought.

For Kant, logical possibility concerns *concepts*, whereas real possibility concerns *things*. Laws of thinking will determine whether a concept is well-formed, but to be well-formed is not yet to achieve objective validity. Conditions on possible experience determine, in addition, what kinds of things could be encountered in the empirical world, what concepts could be instantiated. As such, those conditions on possible experience will determine what *things* are possible, and hence ground real possibilities. As the title of the *Postulates* section suggests, real possibility concerns *empirical thinking*, not thinking in general.

This emerging distinction between thought and cognition is at the heart of the distinction between logical and real possibility.

> To **cognize** an object, it is required that I be able to prove its possibility (whether by the testimony of experience from its actuality or *a priori* through reason). But I can **think** whatever I like, as long as I do not contradict myself, i.e., as long as my concept is a possible thought, even if I cannot give any assurance whether or not there is a corresponding object somewhere within the sum total of all possibilities. But in order to ascribe objective validity to such a concept (real possibility, for the first sort of possibility was merely logical) something more is required. (Bxxvi, footnote)[17]

Logical possibility is to be understood in terms of the conditions for being able to *think* something (laws of thinking); real possibility is to be understood in terms of the conditions for being able to *cognize* something (laws of cognition).

The distinction between what we can or must *think*, and what we can or must *know* or *cognize*, extends throughout Kant's work.

> To **think** of an object and to **cognize** an object are thus not the same. For two components belong to cognition: first, the concepts, through which an object is thought at all (the category), and second, the intuition, through which it is given. (B146)[18]

In summary: cognition involves concepts and intuition. Cognitions are representations which succeed in being about the world. The understanding, as a faculty of thought, contributes concepts and logical form. But in order for a thought to be about an object of experience, an object in the empirical world, it needs to be able to be given an object through intuition—that is, in order to be elevated to the status of a cognition, a thought must be compatible with conditions on possible experience.

Finally, let us return to the postulate of actuality. Whereas possibility is concerned with the mere form of possible experience, actuality differs in being concerned with the *matter* of experience. For some thing to be actual is for it to be 'connected with the material conditions of experience (sensation).' These material conditions

[17] See also A244/B302 and footnote. [18] See also Bxxvi.

of experience pertain to *sensation*: given input through sensibility. There can't be experience—empirical cognition[19]—without empirical intuition, and sensation is required for empirical intuition.[20] The view, at first glance, seems quite plausible. It is simply the view that it is not enough for something to actually exist that it bear certain relations to those features of our cognitive capacities that make it possible for us to experience the world; it must also be part of the world that we experience, independent of our cognitive capacities. For example, looking outside my window I can have a visual experience of a yellow flag with a black 'H' on it. The flag actually exists: it is connected to the material conditions of my having that experience, i.e., it is part of whatever is given to me in sensation. The flag is also compatible with the formal conditions of experience: they allow me to experience the flag, but they don't make it the case that the flag is there.

## 4.3 A Material Condition of Possibility?

Later in the *Critique*, in the *Ideal of Pure Reason*, Kant writes:

> The possibility of objects of sense is a relation of these objects to our thought, in which something (namely, the empirical form) can be thought *a priori*, but what constitutes the material, reality in appearance (corresponding to sensation) has to be given; without that nothing at all could be thought and hence no possibility could be represented. Now an object of sense can be thoroughly determined only if it is compared with all the predicates of appearance, and is represented through them either affirmatively or negatively. But because that which constitutes the thing itself (in appearance), namely the real, has to be given, without which it could not be thought at all, but that in which the real in all appearances is given is the one all-encompassing experience, the material for the possibility of all objects of sense has to be presupposed as given in one sum total; and all possibility of empirical objects, their difference from one another and their thoroughgoing determination, can rest only on the limitation of this sum total. (A581–2/B609–10)

Here Kant appears to have in mind a *material* condition of possibility, but how does that cohere with his account of possibility, earlier in the book, in terms of *formal* conditions? We need to make sense of this if we're to achieve a well-rounded understanding of Kant's views on modality. More importantly, this passage helps us to understand better how our capacity to cognize individuals might contribute to our understanding of modality. To make sense of this passage, though, we first need to consider what comes immediately before.

The action takes place in the *Ideal of Pure Reason*, a chapter of the *Transcendental Dialectic*. Where the *Transcendental Aesthetic* is about the functioning of sensibility, and the *Transcendental Analytic* is about the functioning of the understanding, the

[19] B147.

[20] Note that as well as what we familiarly think of as sensation—through the five senses—for Kant this includes inner sense; our experiences of our own thoughts, feelings, and inner life. See A22–3/B37.

*Transcendental Dialectic* concerns a third cognitive capacity: *reason*. Reason is our capacity for connecting together judgements produced by the understanding (in conjunction with sensibility) in various important ways. Most obviously, reason can make logical (syllogistic) inferences from premises to conclusions. Reason also attempts to reverse this kind of process, looking for the ultimate explanations of things (rather than going 'downwards' from premises to conclusions, reason also tries to go 'upwards', to work out from what premises a given starting point might be a conclusion).

Simplifying greatly, the way in which reason functions gives rise to certain concepts of reason, also called 'ideas'. Kant argues that we inevitably fall under an illusion, taking these ideas—which have an important role to play in organizing our judgements and thoughts into a system—for representations of objects in the world, and that we fall into serious error when we purport to have knowledge on the basis of such an illusion.[21] In the *Ideal of Pure Reason*, Kant gives an account of the origin of our idea of God as arising from the function of reason, and gives a diagnosis of our mistakes in rational theology in terms of the special kind of error we make in mistaking ideas of reason for representations of real things.

First, let us consider Kant's account of the origin of our idea of God. Kant has an account of this idea that doesn't purport to imply the existence of God, but which nevertheless renders it an idea which reason must have.[22] Kant begins by introducing two principles:

> **The principle of determinability**: 'of **every two** contradictorily opposed predicates only one can apply to it [a concept]'. (A571/B599)
>
> **The principle of thoroughgoing determination**: 'among **all possible** predicates of things, insofar as they are compared with their opposites, one must apply to it [the thing]'. (A572/B600)

The principle of determinability is a (general) logical principle, based on the law of non-contradiction. It is logical because it concerns only the form of a cognition about things: they should not be of a contradictory form (see A571/B579).

The latter principle—sometimes called the 'principle of complete determination'—goes beyond the mere (general) logical form of a cognition; it is a synthetic, not a logical, principle. Kant claims that we have to take into account not only the logical form of the

[21] For more on Kant's accounts of transcendental illusion and dialectical error see Grier, *Kant's Doctrine of Transcendental Illusion* (Cambridge: Cambridge University Press, 2001).

[22] The obvious contrast here is with Descartes. According to Descartes, an effect cannot contain more than its cause. The ideas of some things can be caused by us, because they contain no greater reality. For example, I can acquire the idea of a body from my own body. But our idea of God is different. It contains the ideas of perfection and infinitude, neither of which could have their source in us. Only God's existence can explain the origin of our very concept of God. See Descartes, *Meditations on First Philosophy*, translated and edited by J. Cottingham (Cambridge: Cambridge University Press, 1996), 28–31; Third Meditation, 40–5.

concept of the thing, but also the sum of *all* of the possible predicates of things, in order to check whether the thing has one of every opposing pair of such predicates.

> This does not rest merely on the principle of contradiction, for besides considering everything in relation to two contradictorily conflicting predicates, it considers every thing further in relation to **the whole of possibility**, as the sum total of all predicates of things in general; it represents every thing as deriving its own possibility from the share it has in that whole of possibility. The principle of thoroughgoing determination thus deals with the content and not merely the logical form. (A572/B600)

It is enough for the concept of a thing to lack contradiction to satisfy the first principle. For us to grasp the second principle, we must compare the thing to all of the contradictorily opposed pairs of ways things could possibly be, to check that it is indeed determinately one way or another.

Stang dismisses what he calls the 'all possible predicates' argument:

> Several commentators have read Kant as claiming that the principle... is synthetic because it refers to all possible predicates: it says that every object is fully determinate with respect to every possible predicate. However,... [b]y parity of reasoning, a similar argument would also show that the principle of non-contradiction is synthetic, for it states that for all possible predicates $F$ and all possible objects $x$, $\neg(Fx \& \neg Fx)$.[23]

However, whilst it is true that both principles make reference to all possible predicates, importantly different claims are made about them. The principle of determinability says that only one of two contradictorily opposed predicates *can* apply to one (concept of a) thing. Such a principle is a universal rule—referring to *every* two predicates—that can be applied on a case by case basis. For any concept of a thing, it should be free from contradiction. But we are not required to check the concept against every possible predicate. For example, it is enough that the concept *blue carrot* is free from contradiction; we don't need to check it against predicates such as 'viscous' or 'shining'. By contrast, the principle of thoroughgoing determination says that one of every two opposed predicates *must* apply to the thing. This time, we cannot rest content with an incomplete yet internally consistent concept; we must go beyond the concept and consider all possible predicates, working under the assumption that the *thing* is thoroughly determinate with respect to all predicates, and hence that our concept needs to be continually augmented if it is to be at all adequate to the thing.

Stang holds a different view, and argues that the principle is synthetic because it rests on a distinction in transcendental logic—the distinction between negative and infinite judgement—which concerns the content of judgements, and which is ignored by general logic. General logic, that is, fails to distinguish between judgements of the

[23] N. F. Stang, 'Kant on Complete Determination and Infinite Judgment', *British Journal for the History of Philosophy* 20 (2012): 1117–39, pp. 1119–20.

forms 'It is not the case that *a* is *F*' (negative) and '*a* is non-*F*' (infinite). As Stang points out, something's determinately lacking a property does not entail that it determinately has the opposing property, i.e. it doesn't follow from '¬(*a* is *F*)' that '*a* is non-*F*' (*a* might be indeterminate with respect to its being non-*F*). So the claim that everything is determinate, cashed out as the claim that one or other of every two opposite predicates applies to every thing—not merely that every predicate either applies or does not apply—involves drawing on the resources of infinite judgement, which are unavailable to general logic.

Stang helps us to understand the nature of the 'contradictorily opposed predicates' mentioned, but it seems to me that the heart of the matter lies with the 'all possible predicates' argument. This latter argument helps us to understand why a *whole* of possibility is required. Stang's account of this appeals to considerations arising later in Kant's discussion, to do with the sum total of empirical reality. But it should be possible to understand the principle of reason itself, before we move on to discover the empirical version of it.

In order, then, to grasp or apply the principle of thoroughgoing determination we have to consider each thing in relation to '**the whole of possibility**, as the sum total of all predicates of things in general' (A572/B600), the '**sum total of all possibility**' (A573/B601). We must ask ourselves whether, for every possible way something could be, every thing is determinately either that way or determinately not that way. For Kant, this concerns not every pair of *given* contradictory predicates, i.e. the properties we actually experience things as having, but one of every pair of *possible* predicates.

> The proposition **Everything existing is thoroughly determined** signifies not only that of every **given** pair of opposed predicates, but also of every pair of **possible** predicates, one of them must always apply to it. (A573/B601)

As such, the idea of the sum total of all possible predicates is an idea of which we could never experience an instance in the world, i.e. it can't be a concept that applies to objects in the world. It is simply 'too big' for us—we can only experience the totality of how things are given to us; not the totality of how things could be.[24]

> What it means is that in order to cognize a thing completely one has to cognize everything possible and determine the thing through it, whether affirmatively or negatively. Thoroughgoing determination is consequently a concept that we can never exhibit *in concreto* in its totality, and thus it is grounded on an idea which has its seat solely in reason. (A573/B601)

How do we get from here to God? First, this sum total of all possibility is itself a thoroughly determinate concept. Putting together every predicate into a thoroughly

[24] Kant also holds that we can't experience the world as a whole, although reason needs an idea of the world whole. Mistaking this idea for something that we could encounter in experience leads to Kant's famous 'antinomies'. By 'the totality of how things are given to us' I therefore mean something less than the total world whole. What that is should become clearer later in this chapter.

determinate representation of some total thing looks like the concept of an individual object with all those properties.

> [It] becomes the concept of an individual object that is thoroughly determined merely through the idea. (A574/B602)

Kant argues in addition that what he calls 'reality', by which we can understand something like positive properties, is (are) prior to negation. $F$ is prior to $\neg F$.

> Now no one can think a negation determinately without grounding it on the opposed affirmation. The person blind from birth cannot form the least representation of darkness, because he has no representation of light... All concepts of negations are thus derivative, and the realities contain the data, the material, so to speak, or the transcendental content, for the possibility and the thoroughgoing determination of all things. (A575/B603)

So the sum total of possibility boils down to the concept of an individual which has all positive properties, i.e. all realities.

> Thus if the thoroughgoing determination in our reason is grounded on a transcendental substratum, which contains as it were the entire storehouse of material from which all possible predicates of things can be taken, then this substratum is nothing other than the idea of an All of reality (*omnitudo realitatis*). (A575–6/B603–4)

Insofar as the All of reality contains all possible (positive) predicates, from which we can derive all predicates—including the complex and the negative—whatsoever, we can understand all other things in terms of a *limitation* of this All of reality. If every individual is determinate with respect to every possible positive property (it either has the property or has the contradictorily opposed property), and if the sum total of possibility is the sum total of all possible positive properties, then we can consider every individual in terms of what it lacks compared to the sum total, i.e. in terms of a limitation of some of those possible positive properties in the sum total. And for every permutation of limitations, there will be a possible individual described. Not only can we think of all other things in terms of a limitation of this All, but it is also supposed to follow that, therefore, the All of reality is the *original* or *archetypal* being from which all other 'lesser' realities can be derived.

> Thus all the possibility of things... is regarded as derivative, and only that which includes all reality in it is regarded as original. For all negations (which are the sole predicates through which everything else is to be distinguished from the most real being) are mere limitations of a greater and finally of the highest reality; hence they presuppose it, and as regards their content are merely derived from it. (A578/B606)

Finally, then, we 'hypostatise' the idea of this most real being: we take it for the concept of an existent being which grounds the sum total of possibility.[25] This idea of

[25] For details on the move to a ground, see A579/B607.

reason is then, according to Kant, the subject of rational theology, a discipline which purports to gain knowledge of God through reason alone (see A580/B608). But on Kant's view, we make a mistake when we take the idea of a sum total of possibility—a presupposition of our thinking of things in general—for the concept of an individual about which we can learn facts, e.g. that such a thing exists. Our concept of God as the most real being results from the idea of an individual which grounds the sum total of all possibility. Such an idea is an inevitable consequence of how reason functions, but one which presents us with an 'illusion'. Such an illusion in and of itself is not problematic, so long as we recognize it as such and treat it accordingly.[26] But to 'hypostatize' this idea of reason—to turn it into the idea of an individual being—is mistaken, and leads to error.

Note that there is an additional error here. First, it is assumed that a thoroughly determinate concept involving all positive predicates would succeed in picking out an *individual*. But that is to assume that there couldn't be two qualitatively identical yet distinct such beings, i.e. it is to assume the identity of indiscernibles. There has been no role for intuition to play in the account so far, so no wonder we fall into such error. Second, this error is compounded by understanding possible individuals in terms of limitations of the sum total of possibility. Again, such determinate concepts, though limited, would not—at least in most cases—be guaranteed to pick out a unique possible individual, but only at best a possible kind of individual. So the error is in fact twofold: not only mistaking a necessary idea of reason for the concept of a being, but also thinking that a limitation of the sum total would give you possible *individuals*.

Finally, then, Kant explains the kernel of truth in this. This brings us back to the passage from the beginning of this section. I will work through the passage (and what comes immediately before and after) to explain what is going on. In much of my reading of the passage I follow Longuenesse, although we differ in how we situate it in Kant's wider system.[27]

First, Kant asks why reason makes the mistake of coming to think that a single being, God, is the ground of all possibilities of things?

[26] Kant compares what he calls 'transcendental illusion' to the example that the moon looks larger as it rises just above the horizon, compared to how it looks when high in the sky. We can't help but see the moon this way, but we can recognize that it is an optical illusion, and adjust our astronomical calculations accordingly: '[This is] an **illusion** that cannot be avoided at all... just as little as the astronomer can prevent the rising moon from appearing larger to him, even when he is not deceived by this illusion.' (A297/B354)

[27] In *Kant and the Capacity to Judge*, translated by Charles T. Wolfe (Princeton: Princeton University Press, 1998), Longuenesse is working within a framework of understanding the categories in terms of the concepts of reflection (discussed in the *Amphiboly* section), and thereby understanding modality in terms of the concepts of matter and form. By contrast, I am attempting to situate our understanding of modality, and this passage, within a framework which puts the emphasis on a distinction between thinking and cognizing, and the conditions of each. My aim in this work is to find a way to integrate Kant's comments on the sum total of empirical reality with a wider understanding of his account of modality. To do full justice to Longuenesse's views here would require too great a diversion from the main discussion. Interested readers should also consult Longuenesse, 'The Transcendental Ideal, and the Unity of the Critical System' in *Kant and the Human Standpoint* (Cambridge: Cambridge University Press, 2005), 211–35.

How does reason come to regard all the possibility of things as derived from a single possibility, namely that of the highest reality, and even to presuppose these possibilities as contained in a particular original being? (A581/B609)

Kant invites us to look back to his account of cognition and the understanding.

The answer suggests itself on the basis of the discussions of the Transcendental Analytic themselves. The possibility of objects of sense is a relation of these objects to our thought, in which something (namely, the empirical form) can be thought *a priori*, but what constitutes the material, reality in appearance (corresponding to sensation) has to be given; without that nothing at all could be thought and hence no possibility could be represented. (A581/B609)

He reprises the account of real possibility in terms of formal conditions of experience—an object of sense, i.e. the kind of object we can have experience of, is possible if the 'empirical form' of it can be thought a priori, i.e. if the concept of the object is compatible with formal conditions of cognition. But Kant also emphasizes the importance of empirical intuition in cognition: we need given material in order to have representations of individual objects.

Is the intuition requirement too strong? If cognition really requires an intuition, then how can we have any cognition of mere real possibility (e.g. that there could be blue carrots, given that the whole point is that there is no actual intuition of such things)? In answer, this is too strong a reading of the intuition requirement. It must be *possible* for a thought to be given an object in intuition for it to be a cognition of a real possibility, but that just amounts to the claim that the thought be compatible with the formal conditions of possible experience, which is to say it is (at least) compatible with the conditions under which an object could be given to us in intuition. So cognition doesn't always require an intuition, but only that it be 'intuitable'. The kind of cases where intuition *is* genuinely required, are cases of *actual* cognition of things.

The passage continues:

Now an object of sense can be thoroughly determined only if it is compared with all the predicates of appearance, and is represented through them either affirmatively or negatively. (A581/B609)

This appears to be an application of the principle of thoroughgoing determination, but in a restricted form. It is restricted (1) to objects of sense, i.e. the objects of which we can have experience, and (2) to predicates of appearance, i.e. the predicates applicable to objects of sense. At this point we might wonder whether it is legitimate for Kant to assume that there is no indeterminacy in the empirical world, but for now that's an assumption I'm willing to grant him.[28]

[28] In 'Kant on Complete Determination and Infinite Judgment', Stang argues that there is room for indeterminacy of empirical objects within Kant's transcendental idealism. Details aside, my concern here is the role that Kant's apparent claim to determinacy here plays in the wider context of his understanding of modality, rather than whether it is properly consistent for him to make that claim.

Next, Kant appears to reiterate the intuition requirement:

But because that which constitutes the thing itself (in appearance), namely the real, has to be given, without which it could not be thought at all... (A581/B609)

and moves on to a claim about the *unity* of what is thereby given:

but that in which the real in all appearances is given is the one all-encompassing experience... (A581–2/B609–10)

It is an important claim underlying much of the *Critique*'s argumentation that our experience is unified. But, on Kant's view, most aspects of the unity of experience arise through synthesis of the manifold of intuition, which is carried out by the understanding (through the transcendental unity of apperception and use of the categories).[29] This all comes downstream of being given the matter of experience (the manifold of intuition which is to be synthesized). The only unity that could be relevant for 'the real in appearances' as 'given in the one all-encompassing experience' could be the unity already present in given intuition, i.e. the unity of space and time. Indeed, Longuenesse describes the sum total that emerges from this line of thought as the 'whole of reality given in space and time'.[30]

Stang presents a very different interpretation of 'the one all-encompassing experience'.[31] He takes as his starting point a passage from the *Antinomies* chapter, in which genuine experience of the world is distinguished from dreams in terms of the former being 'connected up according to empirical laws in one experience' (A493/B521). Thus Stang claims that 'the "one" all-encompassing experience is grounded in experiences that cohere with each other' (p. 1130), and is 'the lawful representation of the empirical world that is maximally systematic and maximally justified by the totality of sensory states (perceptions) of human subjects' (p. 1132). Stang also marshals a number of passages connecting actuality and connection to sensation through empirical laws in aid of his interpretation.[32]

To properly adjudicate between Stang's interpretation in terms of empirical laws, and my preferred interpretation in terms of the unity of space-time, would require a lengthy, detailed comparison of the various assumptions and motivations underlying our different approaches. But here are some reasons for favouring my approach—bearing in mind that Stang remains in the background as an interesting alternative. It is true that Kant does appeal to empirical laws in his explanation of actuality.[33] But it

[29] See the Transcendental Deduction, especially B129–169.

[30] Longuenesse, 'The Transcendental Ideal', p. 229.

[31] See Stang, 'Kant on Complete Determination and Infinite Judgement' for quoted material in this paragraph, except where otherwise stated.

[32] E.g. B279, A218/B279.

[33] For example, in a particularly strong statement, he writes: 'wherever perception and whatever is appended to it in accordance with empirical laws reaches, there too reaches our cognition of the existence of things. If we do not begin with experience, or proceed in accordance with laws of the empirical connection of appearances, then we are only making a vain display of wanting to discover or research the existence of any thing.' (A226/B273–4)

seems plausible that considerations of this kind come downstream from his concern in the relevant passages in the *Transcendental Ideal*. Here, his concern is with the requirement that the material of experience be given to us, and how it is given to us, rather than with the conceptualization of that given material. The notion of being *given* plays a pivotal role in the passage, occurring throughout, and linking the argumentative steps. According to Kant, insofar as appearances are given to us, they are given in space and time, which provides us with resources for individuating objects (as argued in the *Amphiboly*). Of course, more is required before we achieve full experience of the phenomenal world (application of the categories), and there will be more to be said about how spatiotemporal individuation works given additional conceptual resources, but nothing I've claimed here rules that out. It is just that my preferred reading focuses on an 'earlier' stage of Kant's account of cognition. It thus also appeals to fewer supporting claims, i.e. claims primarily about the form of sensibility, in which the material of experience is given to us, and leaving aside additional claims about how that is transformed into experience by the understanding.

From the claim that the real in appearances—empirical intuition—is given in a unity, Kant infers that this real is therefore given in one sum total:

> the material for the possibility of all objects of sense has to be presupposed as given in one sum total; (A582/B610)

He then repeats the move from a sum total of possibility, to understanding particular possibilities in terms of limitation of that sum total:

> and all possibility of empirical objects, their difference from one another and their thoroughgoing determination, can rest only on the limitation of this sum total. (A582/B610)

But how does this limitation move now work? Above I noted that even a concept that is thoroughly determinate with respect to every possible predicate could not guarantee reference of a particular individual, because there could be two distinct intuitions of objects falling under the concept. Why should restricting the principle of thoroughgoing determination to fewer predicates solve this problem?

The answer turns on the importantly different character of the sum total in which the real of appearances is given, and the sum total of all possibility. The sum total of all possibility, arising from the functioning of reason, collects together all possible predicates of *things in general*, whether they be predicates applicable to things we could experience or not, and hence whether or not every limitation produces something that could be experienced by us. The alternative sum total differs insofar as it only collects together predicates of *appearances*. Such a restriction is not *ad hoc*. The restriction arises because we are now considering a *given* sum total, not a sum total 'determined merely through the idea' (A574/B602). In experience we are given material for experience—a sum total of empirical reality. The proposal is that through limitation of this sum total we can make sense of individuals. But what is that sum total? It is the sum total of the real given to us in empirical intuition, unified by the form of space and time. And that *just is* what Kant claimed is required for us to have genuine thoughts about objects: we need intuitions, and we can only have intuitions in the forms of space and time. The very possibility of our picking out an individual empirical object, and distinguishing it from any other, depends upon our capacity for intuition, and in turn, on difference of spatiotemporal location.

So far so good. But the resulting view appears to have left something like the principle of thoroughgoing determination far behind. How, then, can we understand Kant's apparent appeal to such a principle when he writes that the 'object of sense can be thoroughly determined only if it is compared with all the predicates of appearance, and is represented through them either affirmatively or negatively'? My suggestion is that by 'predicates of appearance' Kant means those predicates which are properly only of appearances, i.e. properly only of 'the undetermined object of an empirical intuition' (A19–20/B34). Before any conceptual determination takes place, the only predicates that already apply to the objects of empirical intuition are those that are already written into the form of intuitions—spatiotemporal predicates. So we end up with the simple, plausible view, which is also consistent with Kant's wider account of our cognitive capacities, that an object of sense can be thoroughly determined—and properly individuated—only if it is compared with *all spatiotemporal predicates* and is represented through them either affirmatively or negatively. In other words, objects of sense need determinate spatiotemporal locations.

So now we have two related principles:

> [P1] 'among **all possible** predicates of **things**, insofar as they are compared with their opposites, one must apply to it [the thing].' (A572/B600)
>
> [P2] among all spatiotemporal predicates of objects of sense, insofar as they are compared with their opposites, one must apply to it [the object of sense].

[P1] applies to 'things in general', and [P2] is modified to apply to 'objects of sense', taking into account the conditions under which something can be an object *for us*. Appealing to the distinction I drew above, we might say that [P1] is a principle for thinking, and only [P2] is a principle for cognizing. Hence only [P2] can tell us anything about the real possibility of things. [P1] goes beyond general logical considerations, as is brought out by Stang's discussion of the distinction between infinite and negative judgement, which appears only in transcendental logic.[34] But [P1] still falls short of being a condition of cognition, rather than thinking. The infinite/negative judgement distinction is still only one of judgement form, and has not yet been modified to take account of the conditions under which we can be given objects in experience. Hence, we are still in the realm of laws of thinking, rather than of cognizing.

[34] Stang, 'Kant on Complete Determination and Infinite Judgment'.

Kant rounds off the argument as follows.

Now in fact no other objects except those of sense can be given to us, and they can be given nowhere except in the context of a possible experience; consequently nothing is an object **for us** unless it presupposes the sum total of all empirical reality as condition of its possibility. (A582/B610)

For Kant, the only objects that we can have experience of are those that are given to us through sensibility. Kant has argued that a presupposition of their being thus given is this sum total of empirical reality. Hence, 'nothing is an object for us unless it presupposes the sum total of all empirical reality as condition of its possibility'.

The passage finally ends with a diagnosis of where we go wrong (when we end up with the idea of God, rather than the idea of this sum total of empirical reality).

In accordance with a natural illusion, we regard as a principle that must hold of all things in general that which properly holds only of those which are given as objects of our senses. Consequently, through the omission of this limitation we will take the empirical principle of our concepts of the possibility of things as appearances to be a transcendental principle of the possibility of things in general. (A582/B610)

He argues that we mistake *the sum of all empirical reality as a presupposition of objects for us*, for *the sum of all possibility as a presupposition of things in general.* If we don't consider the conditions under which something can be an object for us, we yield [P1]. We get the general idea that a 'thing' must be thoroughly determinate with respect to any property, but this is not yet applied to take account of the kinds of things we could encounter. This principle still forms an important part of how we think—we think of individuals as being determinate—but it can lead us into error if we forget that this principle can't tell us anything about the kinds of objects we can experience without being suitably constrained. Once we add in the consideration that we can only experience objects insofar as they are given to us in sensibility (that we can only have experience of appearances, not things in themselves), we get a different principle, [P2]. Hence Longuenesse writes:

In the Transcendental Ideal, Kant argued that reason, by an unavoidable illusion, forms the idea of a *totum realitatis*, individuated by an *ens realissimum*, as the condition of the complete determination of individual things in general. But as given to us, things are completely determined, i.e. individuated, only insofar as they are empirically given in space and time. The *totum realitatis* we do have to presuppose as the given condition of their complete determination is thus the... whole of reality given in space and time. However, making this critical point was not putting an end to the purely rational idea of a whole of reality. Rather, it was constraining us to take it for what it is: a mere thought, without an object.[35]

[35] Longuenesse, 'The Transcendental Ideal', p. 229.

## 4.4 Supplementing the Postulate of Possibility

The upshot of this is that we can supplement the account of real possibility given in the *Postulates*.

> Whatever agrees with the formal conditions of experience (in accordance with intuition and concepts) is **possible**. (A218/B265)

What Kant brings out in the *Transcendental Ideal* is that the postulate of possibility can be accused of ignoring one crucial side of Kant's account of cognition.

> Thoughts without content are empty, intuitions without concepts are blind. (A51/B75)

*Nothing could be an object for us unless it was given to us in a sum total of empirical reality*, hence as well as compatibility with formal conditions on experience, the real possibility of something also presupposes such a sum total in which it could be given, and in which it could be individuated.[36] The sum total of empirical reality, in space and time, in which alone objects for us can be individuated, i.e. completely spatio-temporally determined, is therefore a condition for anything being possible at all.

In preparation for introducing the sum total of empirical reality, Kant referred back to his account of possibility in the *Postulates*: 'The answer suggests itself on the basis of the discussions of the Transcendental Analytic themselves...' (A581/B609). But he then drew attention to the other key aspect of the possibility of any experience of an object: we need the material of experience to be *given*, as well as meeting formal conditions. We need sensible intuition as well as the formal aspects of experience, in order to have genuine experience of objects (i.e. empirical cognitions). This brings forth the idea that the formal conditions of sensibility don't give us the whole picture of real possibility: we need to recognize our need for the material of sensibility, given in empirical intuitions, as well. Without any material for experience, there would be no objects of experience at all.

Again, recall the contrast I drew between thinking and cognizing, as connected to logical and real possibility. Real possibility concerns conditions for cognition. Cognition requires sensibility as well as the conceptual resources of the understanding, and so the whole of empirical reality given through sensibility is included as a condition of cognition, and hence as a presupposition of any real possibility. By contrast, the sum total of possibility concerns only how we think (broadly construed, to include the forms of judgement), without regard for how objects are given to us in sensibility. As such, it may provide a guide for thinking, and hence for logical possibility, but not for real possibility. Indeed, at the end of the passage above Longuenesse remarks that the *ens realissimum* 'was constraining us to take it for what it is: a mere thought, without an object',[37] as something that is *thinkable*, but not *cognizable*.

[36] In 'Kant on Complete Determination and Infinite Judgment' Stang also writes: 'Kant's claim in the Ideal draws out a consequence from the Postulates definition: any possible object of experience has both a form and a matter' (p. 1128). But it should be clear by now that we reach a similar conclusion through rather different routes.

[37] Longuenesse, 'The Transcendental Ideal', p. 229.

If it is correct that our understanding of real possibility must not ignore material conditions, why then does Kant not discuss this in the *Postulates*, but in fact wait until more than 300 pages have passed to add something in the *Transcendental Ideal*? In fact, Kant does appear to anticipate this in the *Postulates* where he writes:

> Whether the field of possibility is greater than the field that contains everything actual, and whether the latter is in turn greater than the set of that which is necessary, are proper questions, and can, to be sure, be solved synthetically, *though they also fall under the jurisdiction of reason alone*; for they mean to ask whether all things, as appearances, *belong together in the sum total* and the context of a single experience . . . or whether my perceptions could belong to more than one possible experience . . . *Whether other perceptions than those which in general belong to our entire possible experience and therefore an entirely different field of matter can obtain cannot be decided by the understanding, which has to do only with the synthesis of that which is given.*
> (A230–1/B282–3, my emphasis)[38]

The understanding and its principles—which include the postulates of modality—concern only how given matter of experience is synthesized, and cannot tell us about the matter itself. If we want to consider questions about this matter then, and how it relates to possibility (e.g. whether it could have been different), we must turn to a different cognitive capacity, reason. In particular, a principle of the understanding, such as the postulate of possibility, can only tell us rules for synthesizing the given matter of experience, and can tell us nothing about the sum total in which that matter is given. Although there is much to learn about sensibility earlier in the *Critique*—that its pure forms are space and time,[39] and how space and time relate to the categories[40]—some issues go beyond this. To claim that for there to be empirical objects *something* must be given, and then to consider whether different sum totals of matter could have been given, is to go beyond our knowledge of things as they appear to us, and hence beyond the limits of cognition, into the kind of speculation in which reason can partake, but only at the peril of error. No wonder, then, that a fuller discussion of how the sum total of the matter of experience relates to real possibility is postponed until the part of the *Critique* which deals with issues of reason, not the understanding.

I stated at the outset of this chapter that we should be interested in Kant's views on modality because we might learn something about modality more generally, and not just about the history of philosophy. What, then, can we learn from my account of how to integrate a puzzling passage about the sum total of all empirical reality into Kant's account of modality? The overall picture I've tried to sketch here is one in which possibility is to be understood in terms of our capacity for cognition, i.e. our capacity to have thoughts genuinely about objects. Real possibility—what might nowadays be called metaphysical possibility—is connected to conditions for

[38] There are alternative interpretations of this passage, which I do not have space to properly consider here. See, for example, Abaci, 'The Coextensiveness Thesis'.

[39] A22/B36.

[40] E.g. in B150–3 and A137–47/B176–87.

cognition, whereas logical possibility is connected to conditions for mere thought. More particularly, Kant claims that our capacity for cognition breaks down into the cooperation of two distinct capacities, sensibility and understanding. I have argued in this paper that we should understand Kant as giving an account of real possibility in terms of the conditions under which cognition is possible, and moreover, that the puzzling passage brings out the fact that we should not neglect to take account of the sensible material conditions of cognition, as well as the conceptual and formal conditions.[41] This suggests that *if* we can defend a distinction similar to that between thinking and cognizing, and *if* we can defend a distinction similar to that between sensibility and understanding, then we will have the ingredients for developing a Kantian account of modality. To assess whether these distinctions can indeed be defended is a project for another time, but it seems to me that this is worth pursuing. Rather than looking into the hidden essences of objects, or into the pluriverse of non-actual worlds, or into the kinds of attitudes we happen to have, we should instead investigate the structure of our very capacity to have thoughts which succeed in being about objects to uncover the nature of possibility.

## Bibliography


Abaci, U., forthcoming. 'The Coextensiveness Thesis and Kant's Modal Agnosticism in the "Postulates"', *European Journal of Philosophy* 24:1 (March 2016): 129–58.

Baldwin, T., 'The Inaugural Address: Kantian Modality', *Proceedings of the Aristotelian Society*, Supplementary Volumes 76 (2002): 1–24.

Chignell, A., 'Kant, Modality, and the Most Real Being', *Archiv für Geschichte der Philosophie* 91 (2009): 157–92.

Chignell, A., 'Kant, Real Possibility, and the Threat of Spinoza', *Mind* 121 (2012): 635–75.

Chignell, A., 'Kant and the "Monstrous" Ground of Possibility', *Kantian Review* 19:1 (2014): 53–69.

Descartes, R., *Meditations on First Philosophy*, translated and edited by J. Cottingham (Cambridge: Cambridge University Press, 1996).

Fisher, M. and Watkins, E., 'Kant on the Material Ground of Possibility: From *The Only Possible Argument* to *The Critique of Pure Reason*', *Review of Metaphysics* 52 (1998): 369.

Grier, M., *Kant's Doctrine of Transcendental Illusion* (Cambridge: Cambridge University Press, 2001).

Kant, I., *Lectures on Logic*, translated and edited by J. Michael Young (New York: Cambridge University Press, 1992).

Kant, I., 'Metaphysik L2' in *Lectures in Metaphysics*, edited by K. Ameriks and S. Naragon (Cambridge: Cambridge University Press, 1997), 299–354.

Kant, I., *Critique of Pure Reason*, translated and edited by P. Guyer and Allen W. Wood (New York: Cambridge University Press, 1998).


[41] Where formal conditions include the formal conditions for sensibility as well as understanding.

Leech, J., 'Kant's Modalities of Judgment', *European Journal of Philosophy* 20 (2012): 260–84.

Leech, J., 'Making Modal Distinctions: Kant on the possible, the actual, and the intuitive understanding', *Kantian Review* 19:3 (2014): 339–65.

Longuenesse, B., *Kant and the Capacity to Judge*, translated by Charles T. Wolfe (Princeton: Princeton University Press, 1998).

Longuenesse, B., 'The Transcendental Ideal, and the Unity of the Critical System' in *Kant on the Human Standpoint* (Cambridge: Cambridge University Press, 2005), 211–35.

Stang, N. F., 'Kant's Possibility Proof', *History of Philosophy Quarterly* 27 (2010): 275–99.

Stang, N. F., 'Did Kant Conflate the Necessary and the A Priori?', *Noûs* 45 (2011): 443–71.

Stang, N. F., 'Kant on Complete Determination and Infinite Judgement', *British Journal for the History of Philosophy* 20 (2012): 1117–39.

Stang, N. F., *Kant's Modal Metaphysics*, Oxford University Press (2016).

# 5

# Hegel's Expressivist Modal Realism

*Christopher Yeomans*

In attempting to understand Hegel's basic position on the modalities, the central thing to grasp is the precise nature of his modal realism. Hegel is adamant that possibility be understood as an actuality in its own right, and furthermore endorses a convergence of the modalities in which what is actual is equally possible and necessary. But Hegel gives an interpretation to these apparently realist claims that invokes neither possible worlds nor cognitive faculties, nor the fullness of time in which what is merely possible now is necessarily actual at some other time. His realism is rather an *expressivist* realism, or a realism of *manifestation*.[1] The actuality of the world is understood as the process through which the nature of the world is constantly expressing or manifesting itself, and so the modalities are taken to describe aspects of that process. Through necessity, with possibility and in actuality, in the unity of their interconnection, the world is manifested as what it is, i.e., in its honour and glory as absolute. And yet this suggestion of pantheism must immediately be qualified: though Hegel treats the modalities as fundamental aspects of the *objectivity* of the world, he holds that their traditional metaphysical articulations through notions of substance and causation cannot do justice to them. Rather, he holds that the world's modal character can only fully be articulated *subjectively*, specifically in judgements. What this means, and how it is compatible with his modal realism, are the issues with which this chapter concludes.

## 5.1 Modal Realism

The problem of interpreting Hegel's modal realism presses itself upon us from the moment we consult his most explicit discussion of modality, in the *Science of Logic*. There we find claims such as the following: '[R]eal possibility is itself *immediate*

[1] Martin Kusch and Juha Manninen, 'Hegel on Modalities and Monadology' in *Modern Modalities*, edited by S. Knuuttila (Dordrecht: Springer, 1988), 109–77, p. 131.

*concrete existence*... because this determination pertains to it by the very fact of being *real* possibility. The real possibility of a fact is therefore the immediately existent manifoldness of circumstances that refer to it' (WL 11.386).[2] Several important things must be noted in this connection. First, for Hegel the modal notions are not exclusively characteristics of propositions or judgements as distinct from the events, entities or other parts of the world to which such propositions and judgements refer. So there is no distinction made between, e.g., modal propositions and their truth-makers. Or, to be more precise, in his initial and most comprehensive discussion of the modalities, there is not yet any distinction made between objectivity and subjectivity. In the final section of this chapter we will turn to his discussion of judgements of necessity, but for now we will simplify things a bit by taking Hegel's initial presentation at face value as a conception of the modality of any kind of fact at all. On this initial take, modality is a comprehensive phenomenon, characterizing everything that could become an object of thought in any way.

Second, Hegel's claim here is not that the possibility of some (actual) fact is its actuality somewhere else, e.g., in another possible world or at another point in time. Instead the claim is that the possibility of a fact is the actuality *of another fact*, in the same world as the first fact. So, for example, the real possibility of a racecar going fast just is the size of its engine, the ratio of its gears, the grip of its tyres, the topography of the racetrack, and so on—as they are used in going fast, before they are used in going fast, or even after they are used in going fast (assuming that they are not destroyed in the process). We will get to the temporal dimension in a bit; for the moment the important point is the relational nature of modality on this conception. If the possibility of one fact is the actuality of another fact then modality is a way in which facts are embedded in relations with each other, into a concrete network of grounding relations.

Hegel's modal name for this network is 'necessity':

> real possibility, since it has the other moment of actuality within it, is already itself necessity. Hence what is really possible can no longer be otherwise; under the given conditions and circumstances, nothing else can follow... But this necessity is at the same time *relative*.—For it has a *presupposition* from which it begins; it takes its *start* from the *contingent*... [T]his necessity... begins from that unity of the possible and the actual which is not yet reflected into itself—this *presupposing* and the *movement which turns back unto itself* are still separate—or [real] *necessity has not yet determined itself out of itself into contingency.* (WL 11.388)

On the one hand, Hegel puts his realist point in quite traditional language, understanding necessity as the unity of possibility and actuality. And within that language

[2] Citations to Hegel's and Kant's works are as follows: 'WL' = Hegel's *Wissenschaft der Logik*, cited by volume and page number from *Gesammelte Werke*. English translations (sometimes slightly modified) are from George di Giovanni's translation, *The Science of Logic*. 'EL' = Hegel's *Encyclopedia Logic*, vol. 20 of *Gesammelte Werke*, cited by section number. 'KrV' = Kant's *Critique of Pure Reason*, cited by A- and B- edition pagination.

he expresses a thought common to Leibniz and Kant as well—that real necessity is hypothetical or conditional necessity and therefore includes some element of contingency. On the other hand, there is new and quite technical language in the mention of reflection into self, presuppositions and movement out of a starting point and into some other modal category. This gets us to a third point we need to make about Hegel's modal realism, which is that it is *reflective* and even *dynamic* in some sense yet to be specified. Since it is not fundamentally about propositions or judgements it cannot be a matter of reflection *on* objectivity; rather it must be a matter of reflection *in* and *through* objectivity. But on its face it is quite difficult to know what this could mean. The key is to see that for Hegel, reflection is fundamentally a matter of expression or manifestation.

There is no space here to go into the technical matter of reflection in any detail, but discussion of a few features of Hegel's understanding of the term will suffice. First, Hegel develops his conception of reflection as an objective or general phenomenon, rather than a subjective, intentional relation to such objective phenomena. Second, he does so as a way of trying to understand how there could be something like a continuant (though Hegel does not use the term), i.e., how something could remain the same through change or variation. In other work I have described this as Hegel's search for an adequate conception of a locus of responsibility in the widest and most general sense of 'responsibility'.[3] After rejecting various quantitative physical models of such continuity, he turns to the idea that things stay the same in virtue of an active relation to themselves in the changes that they undergo or variable relations in which they are entangled. That is, things stay the same in virtue of first expressing what they are under external influence and in context. This is the notion of reflection as expression (or manifestation—I use these expressions interchangeably) that is operative in Hegel's conception of modality:

> The actual is therefore *manifestation*. It is not drawn into the sphere of *alteration* by its externality, nor is it the *semblance* of itself in *an other*. It just manifests itself, and this means that in its externality, and only in *it* is it *itself*, that is to say, only as a self-differentiating and self-determining movement. (WL 11.380–1)

Herbert Marcuse has quite a nice gloss on this central aspect of Hegel's theory of modality that is worth quoting:

> the actual can transform itself and yet remain the same. It can be destroyed, but then *it* is the one destroyed, and this destruction also 'belongs' to it in a sense. Even when it is completely dependent on it, the actual is in active control of its mode of being-there. It does not allow no matter what to happen to it, but resists certain kinds of occurrences, while offering itself to others.[4]

[3] Christopher Yeomans, *Freedom and Reflection: Hegel and the Logic of Agency* (New York: Oxford University Press, 2011).

[4] Herbert Marcuse, *Hegel's Ontology and the Theory of Historicity* (Cambridge, Mass.: MIT Press, 1987), p. 93.

To put it bluntly and in non-Hegelian terms, modality presents the substantiality of things as grounded in things' manifestation of what they are in relational contexts. That is, the substantiality of things is to be found not in their immunity to external influence but in their ability to utilize such influence as a resource for self-expression.

To sum up the basics so far, we have the notion that actuality is a locus of responsibility that is partially grounded in possibilities that are themselves other actualities in the same world, where those grounding relations (when conceived statically) or processes (when conceived dynamically) are necessity itself. So, for example, the actual state of an army is grounded in the possibilities of its size and discipline relative to other armies, its supply lines, the population's willingness to support war, and so on. Yet the very relations of the possibilities to the actual state are a kind of necessity of the army's nature manifesting itself in certain ways. As the saying goes, you can do anything with bayonets except sit on them.

## 5.2 Rejected Views

Before going deeper into Hegel's views, it may be helpful just briefly to review the other options he forecloses. First of all, at least as far as *real* modality is concerned, he does not see the principle of non-contradiction as fundamental. He thinks of what he calls *formal* (i.e., logical) modality as so defined, but he clearly thinks that considered in itself such a form of modality is relatively insignificant (WL 11.382–3). So whatever his argument for the necessity of the actual turns out to be, it cannot be Christian Wolff's claim that at the time that something is actual it is necessary because its contradictory is impossible.[5] In fact, Hegel's argument for the claim that possibilities must be understood as actualities in their own right depends on the negation of Wolff's claim, i.e., it depends on the claim that the actuality of X as defined by the principle of non-contradiction requires the possibility of not-X.[6]

Second, it is clear that Hegel's view cannot be understood via the semantics of possible worlds. The crucial relational nature of Hegel's conception requires real relations of influence (causal or otherwise) between possibilities and actualities, not just the possibility of access to other possible worlds. Furthermore, as Hegel puts it in another context, 'Philosophy does not waste its time with such empty and other-worldly things [*bloß Jenseitige*]. What philosophy deals with is always something concrete and absolutely present' (EL§94Z).

Third, Hegel's view of real modality cannot be understood along the lines of a statistical conception according to which anything that is a real possibility at one point in time is necessarily actualized at another.[7] For one thing, there is officially no

[5] See Martin Kusch and Juha Manninen, 'Hegel on Modalities and Monadology', and chapter 2 of the present volume for more on Wolff.

[6] Kusch and Manninen, 'Hegel on Modalities and Monadology', pp. 123–4.

[7] As a view originating in Aristotle and continuing into medieval philosophers, see Knuuttila, 'Modal Logic' in *The Cambridge History of Later Medieval Philosophy* (New York: Cambridge University Press, 1982), pp. 344–9. For more on this view, both with respect to Leibniz and Wolff and with respect to the role

room for time in Hegel's discussion of the modalities, and this is essential to Hegel's conception as contrasted with Kant. But more importantly, the relational nature of Hegel's conception is lost on this interpretation, and thus also any sense that we can make of Hegel's distinctive conception of the convergence of the modalities (more on this in a moment). Possibilities are always possible with respect to some other actuality, and vice versa, and there must be a different content in these different possibilities and actualities in order to have articulate relations at all. Merely formal differences or differences in spatial location cannot be fundamental, even if they may also characterize the modal facts when more completely considered. So, for example, that which is a possibility with respect to some actuality is itself an actuality with respect to some other possibility and so on throughout the network of necessity, but the shape of that network must be given through the differences of contents of the nodes as much as the different kinds and directions of their connecting rods, as it were.

Fourth, Hegel's conception cannot be understood as a variation on Kant's theme, both because of the unimportance of time but more generally because of the complete absence of appeal to cognitive faculties of any sort. And yet the differences should not be overstated either. In fact, the complex of similarities and differences will help to introduce the next wrinkle into Hegel's view, which is the distinction between real and absolute modality. As noted above, when it comes to real modality Hegel shares with Kant the view that the actual is relatively necessary, i.e., it is necessary relative to some presuppositions. These material presuppositions distinguish real from formal modality, which for both thinkers is governed by the principle of non-contradiction. In Kant's way of thinking about this idea, it comes out as the notion that a state of a substance is necessary only given some previous state on which it is contingent, and the precise nature of this contingent necessity of everything actual is then specified by causal laws connecting the two (actual) states (KrV A226–8/B279–80). At the comprehensive or absolute limit of the relevant previous states and causal laws, the three modalities become coextensive: the possible is actual because it is causally necessary. As part and parcel of this coextension, Kant denies that there is an absolute sense in which one could take the actual world to be *merely* possible, say by the Leibnizian route of imagining that God could have created another world with a very different form had he so desired. So there can be no sense in which absolute possibility has a broader extension than absolute necessity or actuality on Kantian terms; this would require the possibility of representations belonging to another domain of possible experience not structured by causal laws, and Kant doesn't think we can make any sense of this idea. The intensional differences between the modalities even at this coextensive limit can only be maintained by connecting them with distinctive cognitive abilities: possibility to the formal constraints of the understanding, actuality in addition to sensation, and necessity in addition to the connection of

that it plays in the relation between Hegel's three kinds of modality (formal, real, and absolute), see Kusch and Manninen, 'Hegel on Modalities and Monadology', pp. 111–14 and 137–8.

perceptions via concepts (KrV A234/B286).[8] Hegel makes no reference to cognitive faculties in any of his discussions of modality, but neither does he argue for the coextension of possibility and actuality from the perspective of an intuitive intellect for which the two are one because it gives itself its own objects by thinking about them and thus eliminates any contingency in the fit between its representations and its objects.[9] In fact—and here we come to the heart of the issue from the Hegelian perspective—Hegel argues for the coextension not by eliminating but rather by magnifying contingency:

> It is necessity itself, therefore, that determines itself as *contingency*: in its being it repels itself from itself, in this very repelling has only returned to itself, and in this turning back which is its being has repelled itself from itself. Thus has form pervaded in its realization all of its distinctions; it has made itself transparent and, as *absolute necessity*, is only this simple *self-identity of being in its negation*, or in essence. (WL 11.390)

As in the difference with respect to Wolff on the grounding of possibility in non-contradiction noted above, Hegel's view is diametrically opposed to Kant's on this fundamental point.

## 5.3 Convergence of the Modalities

At this point, then, we must shift focus from the first feature of Hegel's modal realism—i.e., the idea that possibilities are themselves actual—to the second feature, i.e., the convergence of all of the modalities, not just possibility and actuality. The basic question that we have to answer here is how Hegel can maintain the distinctions between the intensions of the modal terms if they have become co-extensive. If the distinction between the modalities is not to be thought along Kantian lines as referring to the relation between experiences and particular cognitive capacities, how is that distinction maintained? After all, Hegel makes quite extensive use of the modalities and the differences between them in other areas of his philosophy (for example, in his analysis of the will), so it cannot be the case that Hegel's analysis of modality simply collapses the distinction—or at least not intentionally so on Hegel's part. It turns out that the key to understanding Hegel on this point is his distinction between formal, real, and absolute modality, and the changing significance of the distinctions between possibility, actuality, and necessity in those three different modal frameworks.

[8] For a recent discussion of and dissent from this common interpretation of Kant see Abaci, 'The Coextensiveness Thesis and Kant's Modal Agnosticism in the "Postulates"', *European Journal of Philosophy* 24/1 (March 2016): 129–58, DOI: 10.1111/ejop.12049.

[9] On this point see Sedgwick, *Hegel's Critique of Kant: From Dichotomy to Identity* (Oxford: Oxford University Press, 2012), and Kenneth R. Westphal, 'Kant, Hegel, and the Fate of "the" Intuitive Intellect' in *The Reception of Kant's Critical Philosophy: Fichte, Schelling, and Hegel*, edited by Sally Sedgwick (New York: Cambridge University Press, 2000), 283–305.

As we noted earlier, for Hegel formal or logical modality is characterized by the principle of noncontradiction. Famously or infamously, Hegel holds that this principle can do much less work even at the logical level than most philosophers have believed, primarily because whether something counts as an outright contradiction or just an opposition or tension depends on background assumptions that do the more important work, relegating contradiction to the status of a secondary technique or argumentative tactic (WL 11.383). With respect to Kant this is (positively) the lesson of the Antinomies and (negatively) the lesson of the failure of the Categorical Imperative to play the paramount role in guiding moral reflection, Hegel thinks. With respect to logic more generally, this is represented by the fact that it is almost always tendentious to characterize an opponent's view as containing both *p* and not-*p*. What Hegel calls real modality is the attempt to include those background assumptions of the real context into the modal characterization itself:

> Formal possibility is immanent reflection only as abstract identity, the absence of contradiction in a something. But when we delve into the determinations, the circumstances, the conditions of a fact in order to discover its possibility, we do not stop at this formal possibility but consider its real possibility. (WL 11.386)

Real modality thus suggests a kind of continuum where placement along that continuum is determined by 'how much' information is considered.[10] I use scare quotes here because there is no simple quantitative metric for what counts as more or less information across the board, though it seems as if in individual cases the distinction is clear enough, and it will turn out that this is the feature of real modality that is most important to Hegel.[11]

Keeping this important qualification in mind, we can profitably introduce the contrast between absolute and real modality via the thought experiment of adding information to any given modal description. The question is one of location: where in the modal complex does the new information go? The natural tendency is to see that information going into background conditions, i.e., into real possibility in Hegel's sense: other actualities that enable and constrain the actual fact at issue. At the limit (e.g., the complete state of the world at a time) one then has a firm foundation to attach relations of necessity (e.g., natural laws), and then one gets Kant's conception of the convergence of modalities in determinism. For the moment, however, let us just focus on this location of the addition in the conditions of possibility. Hegel thinks that this addition only draws the actual and the possible together into the bonds of necessity, and in so doing renders the possible more an internal *feature of* the actual

[10] For a helpful, brief discussion of Kant's revival of the distinction between real and logical modality against the Leibnizians, see Andrew Chignell, 'Real Repugnance and Our Ignorance of Things-in-Themselves: A Lockean Problem in Kant and Hegel', *Internationales Jahrbuch Des Deutschen Idealismus* 7 (2011): 135–59.

[11] The improvement (I hope) of the interpretation of this point in what follows over my earlier suggestions in *Freedom and Reflection* was spurred by helpful criticism by Kenneth R. Westphal.

than an external *condition on* it (WL 11.386–7). In the expressivist register of manifestation, the more something necessarily acts on some other thing the more it looks like an aspect of that thing's nature that is then expressed in the actual state of the thing. So, for example, chicken and egg. The natural temptation is to say that in increasing the information relevant to the modality of the chicken we are adding features to the egg—after all, the chicken is our given starting point, that whose modality is at issue. So we additionally consider not just the existence and outward shape of the egg but the process of development going on inside it. And the temptation is to think that as we add this information to the egg, we make the chicken more necessary. After all, it is not just a superficial play of contingent events that connects the chicken and the egg; rather the egg is such a thing as to lead to a chicken. But this is just Hegel's point: the egg is such a thing as to turn into a chicken, i.e., it is a form or feature of the chicken. So though it at first appeared that we were adding the information to conditions of possibility and then secondarily richening up the necessary connections between those conditions and the actuality, it turns out that in so doing we brought the possibility further into the actuality as a determinate feature of its articulated expression and we brought necessity itself into the actuality as its mode of expressing.[12] That is, the significance of the three modal terms has changed: actuality is the whole pattern of variation or course of development rather than the merely existent states; possibilities are the specific contents of that pattern or phenomena in that course (i.e., what were formerly thought to be actualities) rather than external conditions; and necessity is the structure of that pattern or the force of that development, rather than the inevitability of the pattern or development. These shifts in the significance of the three modalities give us absolute as opposed to real modality.

But the crucial thing is that to get this contrast in modal character one needn't add all of the relevant information, either subjectively or objectively; rather, the crucial thing is the nature of the organizing structure and the way that it shifts with each change. In this respect Hegel's view is fundamentally like Kant's in the Antinomies: we don't actually need to have a whole quantitative continuum in view (e.g., of composition and decomposition) in order to be able to make a principled distinction (e.g., between the composite and the simple) at any particular point (KrV A523–4/B551–2). In the thought experiment, the point is the location of the additional information in the organizing structure and the way that changes the significance of the relations between the different parts of the structure, rather than there being some tipping point on the continuum that we reach by the successive addition of information. At every point at which we add (or subtract) information, the significance of the modal terms changes in this way, so the distinction between real and absolute modality is, like the distinction between formal and real modality,

[12] See also Kusch and Manninen, 'Hegel on Modalities and Monadology', p. 140.

available and applicable at every point in conceptual space. To use another Kantian distinction, formal, real and absolute modality constitute a distributive totality in virtue of such applicability, rather than the collective totality of a continuum of modal completeness.

We can bring out further the difference between real and absolute modality by focusing again on possibility, and specifically on the sense in which possibility is *mere* or *alternate* possibility, i.e., the sense in which what is actual could be otherwise—what Hegel calls contingency. For real modality Hegel understands such contingency as a kind of looseness of fit between the conditions of possibility and actuality, i.e., a sense in which the bonds of necessity are sloppy in virtue of the indefiniteness of their attachment points. Here the sense of the unity of possibility and actuality is given by the notion that a different actuality could have been generated if the attachment points had been slightly different. But for absolute modality the sense of 'actuality' has shifted to the whole pattern of production and so the question of contingency is different. Now the possible and the actual exist in a kind of part-whole relation, since possibility has been dragged further into actuality by the additional information. What sense then can be made of their unity?

Hegel's suggestion is to understand the absolute possibility of the actual as the possibility that it could be either (really) possible or actual (WL 11.389–90). The absolutely actual is a kind of magnification of one of its possibilities, or a way one of its possibilities serves as a lens for focusing the other possibilities, and so internal to the actual is the possibility that other possibilities contained within the actual could have served the same function. The possibility that serves as this focusing lens then places the other possibilities on a kind of axis or continuum of real possibility. That is why 'possibility' is not a pun here: the absolute possibility of the actual consists in the way that one possibility defines the axis of other possibilities, i.e., defines their specific nature as really possible. If the egg develops into the chicken we take the development of the chicken as the possibility that defines the absolute actuality of the whole phenomenon and we see other possibilities (e.g., environmental factors) as resources or conditions for such development. On the axis defined by this continuum the nutritional needs of the fox counts as a real possibility, an influence that may explain certain features of the actuality of the live chicken (e.g., its location within a secured henhouse) but otherwise has its significance coloured if not transmuted by the dominant possibility (chicken development) that becomes identical with actuality writ large. But if a fox eats the egg then the development of the chicken is a resource or condition for the nutritional satisfaction of the fox (which is the relevant absolute actuality). On the axis defined by this continuum the development of the chicken plays the role of a real possibility whose influence explains certain secondary features of the nutrition of the fox (e.g., its timing and location, or precise nutrients ingested). In both cases, the absolute possibility that the actual could be either actual or possible is also reflected in the fact that other existences have the status of real possibility.

But this means that the sense in which the actual is merely possible or contingent has changed between real and absolute modality (WL 11.389–90). In real modality, this contingency is a kind of sloppiness or loose fit between conditions of possibility and the actuality for which they are conditions that renders either one or the other relatively indeterminate (depending on which one is given as the point to attach the necessary connections); but in absolute modality it is precisely the tightness of the fit between all three elements that shows that other tight fits are possible. That is, to return to our example, it is precisely the seamless way in which the nutritional needs of the fox fit into the modal story of the developed chicken that indicates the possibility of those needs playing the more important explanatory role at another time or in another respect. But not only has the significance of the alternativeness or *mere* possibility of the actual changed in absolute modality; of course the nature of the actual itself that is merely possible qua contingent has changed. Now the actuality in question is a more complete phenomenon establishing its own axes of possibility, i.e., its own continua of contrasting tight fits. Each alternate possibility is now not just another possible *following state* of the world but another *expression* of the way that the different features of the situation fit together.

## 5.4 Two Modal Illusions

This distinction between real and absolute modality will help us to get at a unique and surprisingly Kantian *dialectical* feature of Hegel's treatment of modality, namely its diagnosis and treatment of two illusions naturally suggested by the distinction between real and absolute modality. The first is the illusion of modal determinism. This illusion arises out of the procedure of the addition of content to modal characterization, a procedure that itself arises out of our sense that real modality is something different from merely formal or logical modality. The temptation here is to think that there is a maximal addition of content to the conditions of possibility that could then be connected with a separate maximal addition to the lines of necessity to generate a picture of the complete state of the world (either at a time or in its past history) and exceptionless laws of nature that would together entail future events. As it plays a role in the debate about free will, the point is that our actions considered as actual events cannot be within our control but rather must be traced back to separate events and laws over which we have no control. This picture is natural enough to pass without much need of justification in a broad swath of modern and contemporary philosophy, but on Hegel's view it is an illusion brought on by misunderstanding of the differential contribution of the new modal information. As we have already seen, Hegel holds that such an addition undermines a key feature of the illusion, which is the discreteness of the three parts (conditions of possibility, necessary laws, and the actual event they together generate). Attending closer to the actual change brought on by the addition shows, Hegel thinks, that necessity and possibility are just pulled closer into actuality, and thus that the picture of a complete state of the world plus causal laws entailing a discrete consequent event is a conceptual impossibility.

But there is a further consequence here that we only hinted at towards the end of the previous section in the distinction between thinking of alternate possibilities as consequent states (i.e., real actualities) and thinking of them as expressions of the tight fit in the modal network (i.e., absolute actualities). Hegel thinks that if we attend to the differential shift in significance from real to absolute modality, and particularly to the importance of the drawing of more features into the necessity of expression of the absolutely actual, what we see is that we have drawn in more pivot points to alternate expressions, and laid them closer to the heart of the actual. Or, to use the metaphor from the previous section, every time we add a new piece of information and bind it tighter to the actual we add another axis of possibility to intersect with those we had before. This adds another potential expression of the way that the whole fits together as, e.g., both the axis of the bee and the axis of the flower provide an expression of the way that their ecosystem hangs together. Thus, for Hegel one might say that contingency and necessity are directly proportional rather than inversely proportional. The way that this arises from attending to the change in significance of absolute as compared with real modality leads us to the second illusion.

The second illusion complements the first by arising from precisely the recognition that possibility and necessity are drawn into actuality. This is the illusion of monism, i.e., the illusion that such successive additions of information to the modal picture would finally lead to a single *ens realissimum*, the one necessary entity (suggested by the part-whole relation of absolute possibility and actuality above). The illusion comes from thinking that the continuum *generated by* the process of absolute actuality is rather *prior to* that process, or determinable outside that process.[13] Though there is no space here to go into the issue in the detail it deserves, a very brief digression through Kant's argument for the necessity of the existence of the *ens realissimum* can set the stage here. We take our reconstruction of Kant's argument from Brady Bowman:

> All possibility is grounded in antecedently given real determinations as its material element. Now, let *impossibility* be defined as that which cancels all possibility whatsoever. If no real determinations at all existed, then all possibility whatsoever would be cancelled. Therefore, it is impossible that no real determinations at all exist. That whose non-existence is impossible is necessary. Consequently, the real determinations that ground all possibility necessarily exist. Now, real determinations are either affirmative or negative (privative) in character; that is, a thing may be determinate either in respect to the properties it has or in respect to those it lacks. But privations refer essentially to the positive determinations whose absence they denote. Therefore, all determinations whatsoever are ultimately grounded in affirmative

[13] Examples of scholars taken in by this illusion are many. Recent examples include Chignell, 'Real Repugnance and Our Ignorance of Things-in-Themselves', pp. 153–7, and Yeomans, *Freedom and Reflection*, sec. 7.3.

> determinations: call these *realities*, and call the set of all such realities the *omnitudo realitatis*. The *omnitudo realitatis* therefore exists necessarily. The steps Kant takes from this distributive totality of all realities to their collective totality or unity in a unique, supremely real (purely affirmative), indivisible, unchanging, eternal being [i.e., the *ens realissimum*] are subtle. He argues for its uniqueness on the basis of his characterization of it as the *ground* of all possibility...[14]

In relation to this two-step process, Hegel's discussion of absolute modality effects a twofold therapy:

First, the absolute possibility that focuses the actuality as a whole need not be considered the end of the axis or continuum of real possibility that it defines. Quite to the contrary: it will define an axis along which even the absolutely actualized is incompletely actualized, and this is also essential to Hegel's expressivism: every expression, even the most perfect, wears on its face the real possibility of alternative and (in some respects) greater expression. The apparent continuum of real possibility is itself shown to be an expression of absolute possibility, in the same way that the contradiction of formal modality is shown to be generated by the additional information of real modality. So however big or healthy is the chicken that makes it out of the egg, there is always the (real) developmental possibility of a bigger chicken, or one that was healthy in a different way. This mitigates the temptation to generate the *omnitudo realitatis*, since even for each axis there is no complete positive determination.

Second, once one has this sort of schema it is easy to see that every modal fact is an intersection of these different axes, without having to think that in every case one axis has to be dominant, i.e., without having to think of the actuality or manifestation of one as coming at the expense of the other(s), as is suggested by the fox/chicken example. Think, in contrast, of flowers and bees. This is a point about substantiality (loosely construed) that turns out to be essential to Hegel's expressivist realism. This mitigates the temptation to construct out of the *omnitudo realitatis* an *ens realissimum* as the single ground of all possibility. In fact, under the influence of Hegel's diagnosis of absolute modality we are supposed to see that the basic distinction between monism and pluralism cannot be fundamental and reflects an inappropriate hypostatization: 'Absolute necessity is not so much the *necessary* [*das* Notwendige], even less *a* necessary [ein *Notwendiges*], but *necessity*—being simply as reflection' (WL 11.393).

Though there is no space here to go into the issue in any detail, I believe that Hegel's discussion of substance and cause as inadequate articulations of modal realism can be reconstructed as turning on the diagnosis and treatment of these two illusions; the crucial thing to see with respect to that modal realism is that the real itself looks different than we might have expected, and cannot easily be identified with either a single entity or even field of entities. At this point, however, we need

[14] Brady Bowman, *Hegel and the Metaphysics of Absolute Negativity* (Cambridge: Cambridge University Press, 2013), p. 119.

rather to turn to Hegel's more positive suggestion for such an adequate articulation, namely his theory of judgement.

## 5.5 Subjective and Objective Modality

As we said in the first section of this chapter, Hegel thinks of modality as a general feature of everything that could be an object of thought, i.e., of the world as such. Nonetheless Hegel holds that the traditional ways of articulating the nature of absolute modality (substance and cause) are both contrastively objective and decidedly inadequate. And the inadequacy of the objective models for absolute modality motivates the development of better conceptual resources that Hegel himself characterizes as contrastively subjective. This one gets simply from his *Science of Logic*, in which these conceptual resources are specifically discussed. But the sense of the contrast becomes even greater when one looks to the rest of Hegel's work and notes that though many of the categories of the objective logic do a lot of work there (e.g., measure, reflection, form and content, and the modalities themselves), the categories of substance and cause do virtually none. So though the argument of the *Logic* commits Hegel to the claim that, in principle, the latter categories are applicable to all objectivity, in practice he sees them as either insignificant or more trouble than they are worth given the misunderstandings that they suggest. So however we are to characterize Hegel's modal realism, we will have to do justice to his idea that the structure of modality has to be given a subjective modelling in order to secure its general applicability and even its objective significance. The obvious initial question is what 'objective' and 'subjective' mean here.

As a way of entering into this question, consider the fact that Hegel provides explicit discussion of judgements of necessity.[15] The most important feature of this discussion is the importance Hegel attaches to the concept of totality and the way that different totalities can be connected together (WL 12.53). Reaching back to the first section of this chapter, we can connect this theme with the idea that for Hegel, modality articulates substantiality in context and under influence; again, emphasizing that we are using non-Hegelian terms to understand Hegel here, we might say that true substance is never lonely. Or, to reach back to the third section, we can connect this theme with the idea that for Hegel, the absolute modality of a fact involves the intersection of different axes of possibility that are represented by actualities in their own right. To return to our examples, the question is how to model the way that the actuality of the fox or the bee constitutes a possibility of the chicken or the flower, without prejudice to the dominance of one over the other or the possibility of their ontological coexistence, as it were. The presence of this theme in Hegel's discussion of judgements is not surprising, since this feature (articulating the relational nature of totalities) is also the basic advantage that Hegel sees in

[15] There is also a discussion of syllogisms of necessity, but I leave that out of consideration here.

subjective models generally. The point of his discussion of objective modality is to show that objectivity itself has to be understood from this perspective, which then gives him the opening to attempt to show that the best models of this necessary understanding of objectivity are, in fact, subjective. So 'subjective' here means deeper self-organization in the context of other instances and forms of self-organization, whereas 'objective' refers to *relatively* independent forms of self-organization; this is why reciprocal interaction is the very outer limit of the conceptual space of objectivity on its border with subjectivity (and is thus the transitional category between these two sides in Hegel's *Logic*). Spatial metaphors fail us here, but we can think of subjectivity as modelling interpenetrating patterns or processes of self-organization, such as the fox and the chicken or the flower and the bee. Objectively speaking, we try to model the fox and the chicken by thinking of them as substances, or causes and effects, but these notions cannot adequately articulate the idea of intersecting modal axes that we saw already to be contained in absolute modality, Hegel thinks. Subjectivity is what allows us to speak of coexisting and codependent totalities, which looks like an outright contradiction if 'totality' is rendered as 'substance' and 'codependence' is understood causally.

But how does subjectivity do so? Again, there is no space here to go into the question in any great detail, but Hegel thinks that Kant has opened up for us access to the ontological significance of forms of judgement. In particular, the idea of totalities that are both self-standing and essentially related is represented in the connection of subject and predicate in a judgement (WL 12.53–4). At least initially, this possibility of co-self-organization or co-self-determination is represented by the idea that the singular (or individual) subject and universal predicate each become what they are by sharing with each other axes of their own character:

> the start is made from the singular as the first, the immediate, and through the judgement this singular is *raised* to universality, just as, conversely, the universal that exists only *in itself* descends in the singular into existence or becomes a being that exists for itself... These two are one and the same—the positing of singularity in its immanent reflection and of the universal as determinate... Through this determinate universality the subject refers to the outside, is open to the influence of other things and thereby confronts them actively. (WL 12.57)

The judgement, that is, is supposed to model the way in which different absolute actualities can intersect and each constitute an axis of possibility for the other. The universal becoming determinate is supposed to model the magnification of possibility into actuality in such a way that the explanatory force of other possibilities is secured, whereas the reflection of singularity is supposed to model the way that actuality absorbs necessity as its own law or drive rather than external force or compulsion. Now it is, of course, an open question whether the terminology of judgement helps any more with this modelling than other vocabularies—your mileage may vary, as the saying goes—but that is Hegel's view and the whole point of his extensive discussion of different forms of judgement. For the purposes of this chapter we will leave this issue aside and see what progress we can make on our general project of understanding Hegel's apparent modal realism by considering more specifically his treatment of what he calls judgements of necessity.

As Hegel frames the issue, the distinctive nature of the judgement of necessity is to be found in the insight that some judgements express the essentially relational nature of their singular subjects in such a way that they are best understood not as predicating a property to a class of objects, but rather to the singular object in its universal significance: 'the subject, e.g. "all humans", sheds its form determination and "the human being" is what it should say instead' (WL 12.77). This universal singular is the genus. But perhaps more importantly, the judgement of necessity comes in three forms (categorical, hypothetical, and disjunctive) which are precisely the forms of judgement that Kant includes under the general heading of *judgements* of relation, and which then generate the *categories* of relation (substance, cause, and interaction) that Hegel considers to be the *objective* articulations of modality. So built into the architectonic relation of Hegel's discussion to Kant's is the claim that *judgements* of necessity do a better job of articulating the *absolute* modal nature of the world than do the *objective categories* of substance and cause, and it appears that the crucial feature of such judgements that allows them to serve this role is their articulation of the universal singular, i.e., the genus, whether as subject or predicate. So in what follows we will just briefly indicate, for each of the kinds of judgement of necessity, how the invocation of the genus modifies the significance of the judgement from Kant in such a way as to more clearly express the co-determination of totalities.

*Categorical judgements*: Hegel thinks that in the attribution of a genus-predicate to an individual-subject the copula expresses a bond of necessity rather than mere being (WL 12.78). That is, when I judge that a rose is red I merely tack an accident onto a bearer, and the connection remains external. This should remind us of real modality, and Hegel's criticism of the model of substance includes the claim that the only kinds of connections it can articulate are precisely these external ones (WL 11, 396). Since the accidents have no internally necessary connection to the substance they cannot articulate the substance's manifestation of its own nature. But when I judge that the rose is a plant, Hegel thinks, the bond is tighter. Furthermore, as part and parcel of this necessity I place both the individual and the genus on a continuum of genera such that it is inherently possible to pick out another genus that would also have such a tight bond:

> But objective universality also has here only its first *immediate* particularization; on the one hand, therefore, it is itself a determinate genus with respect to which there are higher genera; on the other hand, it is not the most *proximate* genus, that is, its determinateness is not directly the principle of the specific particularity of the subject. But what is *necessary* in it is the *substantial identity* of subject and predicate, in view of which the distinguishing mark of each is only an unessential positedness or even only a name. (WL 12.78)

This is, of course, precisely the connection we saw in absolute modality between tightness of fit and alternate possibilities. And it comes from the magnification of one of the possibilities of the actual, as represented in the specific predicate, which as we saw above was the first feature of absolute modality that Hegel wants to model using the subject-predicate form of the judgement. So whereas the alternate possibilities of real modality are represented by the loose fit between substance and accident (e.g., the rose could have been white), the alternate possibilities of absolute modality are represented by the continuum generated by the tight fit between subject-particular and predicate-genus, a tight fit of necessity that is itself generated by the magnification of one possibility of the subject to serve as the focal point for the subject as such. So the distinctively objective articulation of modality can only model real modality, but the subjective articulation can model absolute modality. But notice that on Hegel's view the advance comes from recognizing that only certain types of predicates can play the role of opening up a continuum; 'plant' will do the trick for the rose but not 'red'.[16] This restriction is what constitutes the distinctively subjective nature of the articulation, and this reminds us that 'subjective' in this context refers to depth of self-organization rather than mental life or a personal perspective.

*Hypothetical judgements*: Interestingly, Hegel has very little to say about subjective organization in hypothetical judgements; or rather, what he says is that hypothetical judgements offer rather little in terms of the modelling of subjective organization. In fact he twice says that such judgements are properly not judgements at all: 'The hypothetical judgement, therefore, has a shape which is more that of a proposition; just as the particular judgement is of indeterminate content, so is the hypothetical of indeterminate form, for the determination of its content does not conform to the relation of subject and predicate' (WL 12.79-80).[17] That is, the hypothetical judgement runs away into too many particular forms—ground and consequence, conditioned and unconditioned, cause and effect—so that the restriction on type of predicate that secured self-organization in the categorical judgement is missing. As it stands, this is a merely negative claim: hypothetical judgements cannot do parallel work to categorical judgements, i.e., they cannot model absolute modality. But if we do look for necessity in the connection between antecedent and consequent of a hypothetical judgement we are thus driven back to consideration of the role of the genus in relation to multiple particulars, and a model for this we find in the disjunctive judgement.[18]

[16] On this point see also Stern, *Hegelian Metaphysics* (Oxford: Oxford University Press, 2009), particularly chapter 12.

[17] Cf. WL 12.55: 'It can also be mentioned in this context that a *proposition* can indeed have a subject and predicate in a grammatical sense without however being a *judgement* for that. The latter requires that the predicate behave with respect to the subject in a relation of conceptual determination, hence as a universal with respect to a particular or singular.'

[18] Westphal argues that this interconnection of the three kinds of judgements of necessity is an element of a different kind of realism in Hegel, namely epistemological realism; see his 'Kant, Hegel, and the Fate of "the" Intuitive Intellect'.

*Disjunctive judgements*: If the categorical judgement advances on objective models by using the subject-predicate form of the judgement to model the magnification of a single possibility into the focal point for an actuality as a whole, the disjunctive judgement is supposed to advance on objective models by using the subject-predicate form of the judgement to model the way that actuality absorbs necessity as its own law or drive rather than external force or compulsion. This comes out particularly in the way that Hegel wants to distinguish the disjunctive judgement, properly so-called, from merely empirical enumeration of species: 'An empirical disjunctive judgement is without necessity; A is either B or C or D, etc., because the species B, C, D, etc., are *found beforehand*; strictly speaking, therefore, there is no question here of an "either or"...' (WL 12.81). The exclusive nature of the disjunction in a proper disjunctive judgement can only be secured if the totality of species 'has its *necessity* in the negative unity of the objective universal which has dissolved singularity within itself and possesses, immanent in it, the simple *principle* of differentiation by which the species are *determined* and connected' (WL 12.81). So the question that confronts us is how it can do so, and it will turn out that a further restriction on the universal (this time of the subject rather than the predicate) will generate the '*proximate genus*'. Such a genus would secure true necessity in the multiple realizations of a principle in a way that neither hypothetical nor empirical disjunctive judgements could do.

The whole orientation of Hegel's problem and solution here come out of the difference between real and absolute modality. Several features of this background are familiar from our earlier discussion. First there is the idea from real modality that the multiple possibilities are just given and contingent background conditions that are, as Hegel puts it here, merely found. Absolute necessity is only modelled if there is a way of seeing how these multiple possibilities in fact derive from the driving force or principle of the absolutely actual itself. In subjective terminology, the correct possibility that is made the focal point is a '*greater* universal' that allows it to encompass the particularity of the different species in the predicate (WL 12.80). But this can only happen if the different possibilities are themselves actualities, i.e., if they are themselves different axes of the development of the genus in its determinate form. In the subjective vocabulary Hegel models this move by talking of two senses of membership in a genus, and emphasizing that in the second sense the universal principle itself is placed alongside its particular realizations:

> Now, because the concept is the universal, the positive as well as the negative totality of the particulars, for that reason it is immediately itself also *one of its disjunctive members*; the *other* member, however, is this universality resolved into *its particularity*, or the determinateness of the concept *as determinateness*, in which the very universality displays itself as totality... In the first instance, the disjunctive judgement has the member of the disjunction in the predicate. But the judgement is itself equally disjoined; its subject and predicate are the members of the disjunction. (WL 12.82–3)

So when we have a true disjunctive judgement the subject term itself generates the totality of particular realizations with respect to the set of which the subject term's own universality is only one realization.[19] Hegel's example here is from Goethe's theory of colours: 'If colour is conceived as the *concrete unity* of light and darkness, then this *genus* has within it the *determinateness* that constitutes the *principle* of its particularization into species [i.e., into the various colours]' (WL 12.83). This principle of determinateness then models the internalization of necessity within the absolutely actual.

This is important because it shows that for Hegel the image of the relation of genus to species is not that of a pyramid, or not *only* that of a pyramid. There is no global image of how all genera and species hang together, since there is no global quantitative continuum of subsumption of more or less particularity under the universal (WL 12.82). In terms of traditional models of the absolute this looks like neither pantheism nor monism. But locally we might say that there is the combination of pyramid and line, where the subject term provides the key for translating between the two; or, to use another analogy, the genus is always *primus inter pares*, e.g., the Holy Roman Emperor with respect to the other German princes. We can say that this is the way in which the primacy of judgement as an articulation of modality gives a positive model for the negative feature of modality that we noted at the end of section 5.4, i.e., that the actual cannot easily be identified with either a single entity or even field of entities. There really are genera and species, Hegel thinks, but it turns out that the objective relations between them are more complicated than we might have originally expected. The species manifest the nature of the genus, but this expressive relation requires precisely that the genus appear alongside them as another possible expression.[20]

## Bibliography

Abaci, U., 'The Coextensiveness Thesis and Kant's Modal Agnosticism in the "Postulates"', *European Journal of Philosophy* 24/1 (March 2016): 129–58, DOI: 10.1111/ejop.12049.

Bowman, B., *Hegel and the Metaphysics of Absolute Negativity* (Cambridge: Cambridge University Press, 2013).

Chignell, A., 'Real Repugnance and Our Ignorance of Things-in-Themselves: A Lockean Problem in Kant and Hegel', *Internationales Jahrbuch Des Deutschen Idealismus* 7 (2011): 135–59.

Franks, P. W., *All or Nothing: Systematicity, Transcendental Arguments, and Skepticism in German Idealism* (Cambridge, MA: Harvard University Press, 2005).

[19] Though there is no room to take up the issue here, this relates to Hegel's discussion of the difference between pure and abstract universality. See my 'Power as Control and the Therapeutic Effects of Hegel's *Logic*', *Hegel Bulletin* 36/1 (2015): 33–52.

[20] Though there is no room to go into the issue here, this suggests a way that the mature Hegel can have a systematic philosophy that responds to the imperatives of both holism and monism without falling prey to Jacobi's objection of nihilism, i.e., the objection that holistic monism makes individuality impossible. See Paul Franks, *All or Nothing: Systematicity, Transcendental Arguments, and Scepticism in German Idealism* (Cambridge, MA: Harvard University Press, 2005).

Hegel, G. W. F., *Gesammelte Werke* (Hamburg: Meiner, 1968).

Hegel, G. W. F., *The Science of Logic*, translated by G. Di Giovanni (Cambridge: Cambridge University Press, 2010).

Kant, I., *Critique of Pure Reason*, translated by P. Guyer and A. W. Wood (Cambridge: Cambridge University Press, 1998).

Knuuttila, S., 'Modal Logic' in *The Cambridge History of Later Medieval Philosophy* (New York: Cambridge University Press, 1982), 342–57.

Kusch, M., and Manninen, J., 'Hegel on Modalities and Monadology' in *Modern Modalities*, edited by S. Knuuttila (Dordrecht: Springer, 1988), 109–77.

Marcuse, H., *Hegel's Ontology and the Theory of Historicity* (Cambridge, MA: MIT Press, 1987).

Sedgwick, S., *Hegel's Critique of Kant: From Dichotomy to Identity* (Oxford: Oxford University Press, 2012).

Stern, R., *Hegelian Metaphysics* (Oxford: Oxford University Press, 2009).

Westphal, K. R., 'Kant, Hegel, and the Fate of "the" Intuitive Intellect' in *The Reception of Kant's Critical Philosophy: Fichte, Schelling, and Hegel*, edited by Sally Sedgwick (New York: Cambridge University Press, 2000), 283–305.

Yeomans, C. L., *Freedom and Reflection: Hegel and the Logic of Agency* (New York: Oxford University Press, 2011).

Yeomans, C. L., 'Power as Control and the Therapeutic Effects of Hegel's Logic', *Hegel Bulletin* 36/1 (2015): 33–52.

# 6

# Russell on Modality

*Thomas Baldwin*

Russell's published writings contain several passages in which he discusses necessity and possibility, but he never published an extended treatment of this topic. He often starts by rejecting the assumption that necessity and possibility are properties of propositions and proposes instead that they are properties of propositional functions. Here is a typical passage, taken from Russell's *Introduction to Mathematical Philosophy*:

Another set of notions as to which philosophy has allowed itself to fall into hopeless confusions through not sufficiently separating propositions and propositional functions are the notions of 'modality': *necessary*, *possible*, and *impossible*. (Sometimes *contingent* or *assertoric* is used instead of *possible*.) The traditional view was that, among true propositions, some were necessary, while others were merely contingent or assertoric; while among false propositions some were impossible, namely, those whose contradictories were necessary, while others merely happened not to be true. In fact, however, there was never any clear account of what was added to truth by the conception of necessity. In the case of propositional functions, the three-fold division is obvious. If '$\phi x$' is an undetermined value of a certain propositional function, it will be *necessary* if the function is always true, *possible* if it is sometimes true, and *impossible* if it is never true.[1]

At first sight, Russell's position here seems clear enough. It is not propositions themselves which are necessary, possible or impossible, but propositional functions, or rather their 'undetermined values', which satisfy these 'notions of "modality"'. Russell's explanation in the final sentence indicates that these modal notions can be thought of as quantifiers, which would explain how the modal notions combine with propositional functions to yield complete propositions. But the content of the resulting proposition depends on the domain of these quantifiers. Russell had used the temporal idioms employed here—'always', 'sometimes', 'never'—in 'On Denoting'[2] in a purely ontological, non-temporal, way, to mean 'in all/some/no cases', and it is

[1] Russell, *Introduction to Mathematical Philosophy* (London: Allen & Unwin, 1919), p. 165.
[2] Russell, 'On Denoting', in *The Collected Papers of Bertrand Russell* vol. 4, edited by A. Urquhart (London: Routledge, 1994), 415–27, p. 416.

reasonable to suppose that his use here is much the same. But what are the 'cases' in question here? One suggestion might be that they are possible worlds, which would give us the familiar view that possibility is truth in some possible world, and so on. As we shall see, however, it is not easy to attribute this view to Russell, though I shall suggest that one way to make sense of his position is to attribute to him the thought that possibilities are logical constructions. But my main aim here is just to present the salient features of Russell's discussions of modality and to discuss how they are best understood. I shall start by looking at some of the early writings which lead up to his only paper explicitly devoted to this topic: his 1905 paper 'Necessity and Possibility' which he did not publish but which is now published in volume 4 of *The Collected Papers of Bertrand Russell*. I shall then move on to discuss Russell's subsequent writings about necessity and possibility, such as the passage quoted above.

## 6.1 Early Writings: 1897–1904

Russell's first discussions of necessity arise in the context of critical discussions of the work of Kant and of Leibniz. He deals first with Kant, in the context of his 1897 *Essay on the Foundations of Geometry*, which was based on his successful Trinity College Fellowship dissertation of 1895. Russell's main aim here was to assess how far Kant's treatment of geometry in his Transcendental Aesthetic remains applicable once geometry is understood as the projective geometry developed by Felix Klein and others to take account of non-Euclidean geometry. So a central question for Russell was how far the thesis that geometry is a priori remains defensible once geometry is understood to be projective geometry, and necessity enters into this question via Kant's thesis that necessity is a mark of the a priori. On this point Russell comments:

> But modern logic has shown that necessary propositions are always in one aspect at least, hypothetical. There may be, and usually is, an implication that the connection, of which necessity is predicated, has some existence, but still, necessity always points beyond itself to a *ground* of necessity, and asserts this ground rather than the actual connection.... But the ground of necessity is, so far as the necessary connection in question can show, a mere fact, a merely categorical judgment. Hence necessity alone is an insufficient criterion of aprioricity.[3]

Russell therefore accepts that 'we must supply the hypothesis or ground, on which alone the necessity holds'[4] before stating that:

> in a fairly advanced science such as Geometry, we can, I think, pretty completely supply the appropriate ground, and establish, within the limits of the isolated science, the distinction between the necessary and the merely assertorical.[5]

[3] Russell, *An Essay on the Foundations of Geometry* (Cambridge: Cambridge University Press, 1897), p. 4. A later comment (p. 58, n. 2) indicates that Russell's reference in the first sentence to 'modern logic' is a reference to the logical theories of Bradley and Bosanquet.

[4] Russell, *An Essay on the Foundations of Geometry*, p. 4.

[5] Russell, *An Essay on the Foundations of Geometry*, p. 4.

This then constitutes the main theme of the subsequent discussion. Following Kant, Russell takes it to be his task to show:

> that experience of diverse but interrelated things demands, as a necessary prerequisite, some sensational or intuitional element, in perception, by which we are led to attribute complexity to objects of perception; that this element, in its isolation may be called a form of externality; and that those properties of this form, if any such can be found, which can be deduced from its mere function of rendering experience of interrelated diversity possible, are to be regarded as *a priori*.[6]

And having completed this task (at least in his own judgement) Russell concludes:

> The Kantian argument—which was correct, if our reasoning has been sound, in asserting that real diversity, in our actual world, could only be known by the help of space—was only mistaken, so far as its purely logical scope extends, in overlooking the possibility of other forms of externality, which could, if they existed, perform the same task with equal efficiency.[7]

Russell's discussion of this 'Kantian argument' is sophisticated. It resembles Strawson's later discussion in *Individuals*, of a 'sound world' whose 'auditory analogue of space'[8] is a 'form of externality' of the kind implied by Russell's Kantian argument, although Russell's focus is of course on the possibilities dealt with in projective geometry. Since Russell's argument concerns the conditions for the possibility of empirical knowledge, the considerations invoked are epistemological and its primary conclusion is likewise epistemological, namely that the axioms of projective geometry, which Russell argues apply to any form of externality, are a priori. But the question for us concerns its implications concerning the necessity of geometry, in particular concerning the significance of any factual '*ground*' of this necessity. Russell accepts that the way in which the Kantian argument for the necessity of a form of externality depends on the requirements of objective experience implies that these requirements constitute a ground for this necessity. So this necessity is only conditional, conditional upon 'the constitution of the *mind*', as he puts it, although this does not qualify the truth that the world exhibits a form of externality and thus satisfies the axioms of projective geometry:

> The *ground* of necessity, we may safely say, arises from the mind; but it by no means follows that the *truth* of what is necessary depends only on the constitution of the mind. Where this is not the case, our conclusion, when a piece of knowledge has been declared *a priori*, can only be: owing to the constitution of the *mind*, experience will be impossible unless the *world* accepts certain adjectives.[9]

So while the truth of the axioms of projective geometry is unconditional, their necessity is conditional: 'For these axioms, and these only, are necessarily true of any world in which experience is possible'.[10]

[6] Russell, *An Essay on the Foundations of Geometry*, p. 62.
[7] Russell, *An Essay on the Foundations of Geometry*, p. 186.
[8] Peter Strawson, *Individuals* (London: Methuen, 1959), pp. 74–5.
[9] Russell, *An Essay on the Foundations of Geometry*, p. 179.
[10] Russell, *An Essay on the Foundations of Geometry*, p. 179.

G. E. Moore discussed Russell's book in a long critical notice in *Mind*,[11] concentrating on Russell's Kantian argument and the conclusion Russell took himself to have established by this argument. Moore had recently been awarded a Prize Fellowship at Trinity on the basis of a dissertation on 'The Metaphysical Basis of Ethics'. This included a long chapter 'On the meaning of "Reason" in Kant' which is an extended critical discussion of Kant's conception of the a priori.[12] So it is no surprise that in his discussion of Russell's book Moore is sharply critical of Russell's Kantian argument, and accuses him of repeating the mistakes which are characteristic of Kant and his followers. In particular, Moore argues that when Russell concludes that 'owing to the constitution of the *mind*, experience will be impossible unless the *world* accepts certain adjectives',[13] his argument, so far from establishing an a priori necessity, is just a claim about the requirements of our own psychology. Russell is here just 'putting the ground of necessity in some psychological fact'.[14]

Moore's discussion fails to do justice to Russell's epistemological arguments, which concern the conditions under which sense-experience provides objective knowledge. But so far from challenging Moore's criticisms of him, Russell wrote to Moore that he agreed with them:

> I had not written to you about your review, because on all important points I agreed with it.
> (18 July 1899)[15]

This letter should not be taken, however, as evidence that it was Moore's review which persuaded Russell that his position was untenable. For he had in fact already changed his mind radically both about the nature of necessity and about the way in which claims about necessity are to be established. The evidence for this change of mind is apparent in Russell's reply to an earlier review of his book by Couturat.[16] In his 1898 reply to Couturat, Russell begins by arguing that it is important to separate 'Philosophy' from 'Psychology' (which includes epistemology), so that questions concerning the nature of necessity are kept separate from questions of how necessary truths are known:

> There is, in general, no proof of the true, and no proof of the necessary. The two ideas are ultimate and unanalysable. Our knowledge cannot furnish proof, because, in order that a

[11] G. E. Moore, 'Critical Notice of B. Russell *An Essay on the Foundations of Geometry*', *Mind* 9 (1899): 397–405.

[12] Consuelo Preti and I discuss Moore's criticisms of Kant in our Editors' Introduction (esp. pp. lvii–lxiv) to Moore's *Early Philosophical Writings* (Cambridge: Cambridge University Press, 2011), which includes Moore's Fellowship dissertations. Moore used parts of this chapter in his paper 'The Nature of Judgment' (1899) which was published in *Mind* 8, shortly before Moore's critical notice of Russell's book.

[13] Russell, *An Essay on the Foundations of Geometry*, p. 179.

[14] Moore, 'Critical Notice of B. Russell *An Essay on the Foundations of Geometry*', p. 399.

[15] See N. Griffin, *Russell's Idealist Apprenticeship* (Oxford: Clarendon Press, 1991), p. 134, n.69.

[16] L. Couturat, 'Review of B. Russell *An Essay on the Foundations of Geometry*', *Revue de métaphysique et de morale* 6 (1898): 354–80.

proposition be known, it is necessary first of all that it be true: an erroneous opinion is not knowledge. Similarly, concerning the necessary: we must perceive necessity if we want to know it, but our perception is not its ground. On the contrary, it is an ultimate fact that a proposition is necessary, and our perception, if it is correct, results from the fact, and not conversely.[17]

Thus Russell has now abandoned the claim which was central to the argument of his book, that 'necessity always points beyond itself to a *ground* of necessity',[18] and which, in the book, turned out to be the way in which 'the constitution of the mind' implies that the world's having a form of externality is necessary for the possibility of objective experience. Instead he now emphasizes the separation of these 'psychological' considerations from philosophical claims concerning what is necessary:

Our psychical nature, except to the extent that it is bound by the a priori laws governing everything that exists, seems to be entirely empirical; this is a given fact, not a necessary truth.

For the necessary there does not exist, as far as I can determine, a universal criterion. We perceive that a proposition is necessary, as we perceive that the sky is blue, and our perception is in no way the ground of the fact, although in both cases it is essential to our knowledge.[19]

And it is evident that we cannot always *prove* necessity; for necessity can be proved only by reference to a consequence of that which is necessary, and this itself presupposes the unproved necessity of logical consequence, so that in the end necessity must be purely and simply perceived.[20]

It is unfortunate that this new position is not explored in any detail by Russell, although, as we shall see, there are traces of it in some other places. I shall refer to it as Russell's 'intuitionist' account of necessity; it is, I think, comparable to the conception of necessity as 'primitive modality' that David Lewis discusses and rejects in *On the Plurality of Worlds*.[21] It is also reminiscent of the account of goodness as simple and 'intuitively known' that Moore presented in his 1898 Fellowship Dissertation.[22] Since Moore did not submit this dissertation until September 1898, whereas Russell had sent off his reply to Couturat in August of the same year, it is not clear whether it is appropriate to regard Russell's new intuitionist account of necessity as evidence of Moore's influence, instead of seeing them as sharing a reaction against the Kantian reasoning with which they were both all too familiar, and adopting instead intuitionist approaches to ethics and to metaphysics. I expect that the latter account of the situation is correct; nonetheless Russell was always content to give the credit to Moore:

It was toward the end of 1898 that Moore and I rebelled against both Kant and Hegel. Moore led the way, but I followed closely in his footsteps.[23]

[17] Russell, 'Are Euclid's Axioms Empirical?', translated by N. Griffin and G. H. Moore in *The Collected Papers of Bertrand Russell* vol. 2, edited by N. Griffin and A. C. Lewis (London: Routledge, 1990), 325–38, p. 333.
[18] Russell, *An Essay on the Foundations of Geometry*, p. 4.
[19] Russell, 'Are Euclid's Axioms Empirical?', p. 333.
[20] Russell, 'Are Euclid's Axioms Empirical?', p. 335.
[21] David Lewis, *On the Plurality of Worlds* (Oxford: Blackwell, 1986), pp. 150ff.
[22] Moore, *Early Philosophical Writings*, pp. 177–8.
[23] Russell, *My Philosophical Development* (London: Allen & Unwin, 1959), p. 54.

An important task which Russell had agreed to undertake soon afterwards in his new role as philosophy lecturer at Trinity College was to provide lectures in Lent Term 1899 on Leibniz (as a replacement for the usual lecturer, McTaggart, who was away at the time). This required Russell to study Leibniz carefully for the first time, and the result of these studies was his next book, *A Critical Exposition of the Philosophy of Leibniz*. Russell was clearly excited and inspired by Leibniz's writings. Yet although in his book he famously argued that logic was central to Leibniz's philosophy, he rejected Leibniz's claim that the necessity of logic and arithmetic arises from the fact that they follow from the 'Law of Contradiction' alone, which, using Kantian vocabulary, he described as the thesis that their necessity is 'analytic'. So, while agreeing with Kant that 'the propositions of Arithmetic, as Kant discovered, are one and all synthetic', Russell held that 'the doctrine of analytic propositions seems wholly mistaken'.[24] According to Russell, even the truths of logic are not analytic, since 'no proposition can follow from [the Law of Contradiction] alone, except the proposition that there is truth'.[25] This position was in line with that which he had advanced in *An Essay on the Foundations of Geometry*, according to which the law of contradiction is 'powerless' by itself. In that book he had associated this claim with a general scepticism about the analytic/synthetic distinction, according to which:

> Every judgment—so modern logic contends—is both synthetic and analytic; it combines parts into a whole, and analyses a whole into parts.[26]

where the reference to 'modern logic' is explained in the following footnote:

> I have stated this doctrine dogmatically, as a proof would require a whole treatise on Logic. I accept the proofs offered by Bradley and Bosanquet, to which the reader is referred.[27]

However, there is no sign of a similar scepticism about the analytic/synthetic distinction in *A Critical Exposition of the Philosophy of Leibniz*; and instead of relying on the contentions of Bradley's 'modern logic', Russell here begins to turn against Bradley's logic, most notably the assumption that 'a subject and a predicate are to be found in every proposition'[28] which he takes to have been a fundamental error of both Leibniz and Bradley. So in place of Bradley's logic, the 'powerlessness' of the Law of Contradiction is elucidated by arguing that what makes an idea such as that of a round square self-contradictory is that it 'involves' both the truth and falsehood of some one judgement and that this involvement rests on the synthetic relation of incompatibility. Thus in the case of the idea of a round square the involvement arises from the fact that 'this is a round square' implies both 'this is round' and 'this has four

[24] Russell, *A Critical Exposition of the Philosophy of Leibniz* (London: Allen & Unwin, 1900) p. 21 and p. 22.

[25] Russell, *A Critical Exposition of the Philosophy of Leibniz*, p. 22.

[26] Russell, *An Essay on the Foundation of Geometry*, p. 58.

[27] Russell, *An Essay on the Foundation of Geometry*, p. 58, n.2.

[28] Russell, *A Critical Exposition of the Philosophy of Leibniz*, p. 12.

angles', which are incompatible because being round 'involves the incompatibility of its constituents with the possession of angles', and, Russell emphasizes:

> But for this synthetic relation of incompatibility, no negative proposition would occur, and therefore there could be no proposition involved which would be directly contradictory to the definition of a square.[29]

Russell's conclusion here, therefore, is that neither arithmetic nor logic follow from the Law of Contradiction alone, and thus that in both cases the truths involved are synthetic. So, he writes:

> it is evident from what has been said already, that if there are to be any necessary propositions at all there must be necessary synthetic propositions. It remains to enquire what we mean by necessity, and what distinction, if any, can be made between the necessary and the contingent.[30]

In answer to the first question—'what we mean by necessity'—Russell now introduces his intuitionist account of necessity: 'It would seem that necessity is ultimate and indefinable.' What is then a little odd is that Russell does not connect this conception of necessity with the synthetic relation of incompatibility he has just been writing about. For, on the face of it, to say that propositions $p$ and $q$ are incompatible is just to say that it is necessary that $p$ and $q$ are not both true. Perhaps Russell took this to be too obvious to merit explicit statement.

However that may be, Russell now turns his attention to the question of the distinction between the necessary and the contingent. He begins by acknowledging that both Leibniz and Kant held that this is a fundamental distinction, such that, for example, 'the propositions of mathematics are necessary, while those asserting particular existence are contingent'.[31] But he then adds a sceptical opinion of his own:

> It may be questioned whether this distinction is tenable, whether, in fact, there is any sense in saying, of a true proposition, that it might have been false.... And it must be confessed that, if *all* propositions are necessary, the notion of necessity is shorn of all its importance.

Russell does not provide any grounds for this scepticism, though he remarks that despite Leibniz's frequent references to the contingency of true propositions concerning what actually exists, his metaphysics implies that they are necessary:

> They themselves, however, have their sufficient reason in God's goodness, which one must suppose metaphysically necessary.[32]

[29] Russell, *A Critical Exposition of the Philosophy of Leibniz*, p. 20 and p. 21.
[30] This and the following quotation: Russell, *A Critical Exposition of the Philosophy of Leibniz*, p. 23.
[31] This and the following quotation: Russell, *A Critical Exposition of the Philosophy of Leibniz*, p. 24.
[32] Russell, *A Critical Exposition of the Philosophy of Leibniz*, p. 39.

This line of thought is not one that Russell himself would have been willing to endorse. But it may be that he had been impressed by Moore's reasoning for a similar conclusion in his 1899 'The Nature of Judgment'. Moore ends his paper with the claim:

That a judgment is universally a necessary combination of concepts, equally necessary whether it be true or false.[33]

Moore's reasoning for this conclusion starts from the claim that the truth or falsehood of a proposition is not contingent upon an external relation such as correspondence between the proposition and any existents. Instead:

A proposition is constituted by any number of concepts, together with a specific relation between them; and according to the nature of this relation the proposition may be either true or false. What kind of relation makes a proposition true, what false, cannot be further defined, but must be immediately recognised.[34]

So the relation which combines the constituent concepts of a proposition itself determines whether the proposition is true or false. Hence the truth of a true proposition is internal to it, and thus necessary. This applies even to a simple existential proposition such as 'Red exists now', which Kant and others would take to be contingent; according to Moore, when I affirm this proposition:

my meaning is that the concept 'red' and the concept 'existence' stand in a specific relation both to one another and to the concept of time.... And this connexion of red and existence with the moment of time I mean by 'now', would seem to be as necessary as any other connexion whatever. If it is true, it is necessarily true, and if false, necessarily false.[35]

I should add that, so far as I am aware, Russell does not explicitly endorse this aspect of Moore's position although he greatly admired 'The Nature of Judgment'. Moore himself changed his mind on this point quite soon, for in his 1900 paper 'Necessity' he argued that 'no proposition is necessary in itself' and proposed instead that the supposed necessity of a proposition consists only in its 'priority' relative to other propositions (I shall say more about Moore's suggestion below).[36]

As we shall see below, in several of his later writings Russell alludes to 'Leibniz's conception of many *possible* worlds', so before leaving *A Critical Exposition of the Philosophy of Leibniz* we should briefly take note of what he says here about this conception. According to Russell, Leibniz's theory is that possible worlds, of which there are an infinite number, consist of individual substances endowed with activity.[37] A substance is possible when its existence does not involve a contradiction; but substances are compossible only where they constitute a possible world, which is a

[33] Moore, 'The Nature of Judgment', *Mind* 8 (1899): 176–93, p. 192.
[34] Moore, 'The Nature of Judgment', p. 180.
[35] Moore, 'The Nature of Judgment', pp. 189–90.
[36] Moore, 'Necessity', *Mind* 9 (1900): 289–304, pp. 302–3.
[37] Russell, *A Critical Exposition of the Philosophy of Leibniz*, p. 68.

world which God might have created, with the result that 'each possible world depends upon certain designs or ends of God proper to itself'.[38] Russell takes this point to introduce the important requirement that for each world there needs to be a sufficient reason for things being as they are, even if it is not self-contradictory to suppose that they are otherwise; and thus that in each world there must be general laws which determine causal connections between substances. So it is through satisfying general causal laws that a plurality of possible substances becomes a possible world whose constituents are compossible. Some of the details of this account are disputable, but it is not important here to decide how far Russell's account of Leibniz is correct. For our present purposes two points are nonetheless worth noting: (i) the fact that Russell does not suggest that one might seek to ground the notion of possibility (or necessity) on that of a possible world; (ii) although Russell accepts that a substance is possible where its existence does not lead to a contradiction, he shows by a reference back to his previous discussion that the question of the compossibility of substances takes one back to the questions which he had dealt with in his earlier discussion of the synthetic relation of incompatibility, and thus, albeit only implicitly, to his discussion of necessity.

Russell's next book was *The Principles of Mathematics*. His main aim here was to vindicate the Leibnizian thesis that 'all mathematics is deduction by logical principles from logical principles'[39] by using the new logical theory which his discovery of Peano's writings had led him to develop. Although he completed a first draft of the book at high speed by the end of 1900, early in 1901 he discovered a fundamental contradiction (now commonly known as 'Russell's paradox') in his theory of classes, which was an essential part of his logical theory. This necessitated substantial revision of the logical theory which took him another year, so that the book was eventually published only in 1903. Despite the fact that Russell took it that he was in the end able to vindicate Leibniz's logicist thesis that mathematics is deducible by logical principles from logical principles, he did not give this result the significance that Frege in 1884 had given to the similar conclusion that 'every proposition of arithmetic [is] a law of logic'. For Frege agreed with Kant that logic is analytic and therefore inferred that 'the laws of arithmetic are analytic judgments'.[40] Russell, by contrast, still maintained that 'logic is just as synthetic as all other kinds of truth', and so for him the vindication of Leibniz's logicism implied that Kant 'rightly perceived that [the propositions] of mathematics are synthetic'.[41]

[38] Russell, *A Critical Exposition of the Philosophy of Leibniz*, p. 67. The passage comes from Leibniz's letter to Arnauld of July 14, 1686 (G. II. 51; see Loemker, *G. W. Leibniz: Philosophical Papers and Letters* (Dordrecht: Reidel, 1969), p. 333).

[39] Russell, *The Principles of Mathematics* (London: Allen & Unwin, 1903), p. 5.

[40] This and the previous quotation: Gottlob Frege, *The Foundations of Arithmetic*, trans. J. L. Austin (Oxford: Blackwell, 1953), p. 99.

[41] Russell, *The Principles of Mathematics*, p. 457. In support of this claim he refers to the discussion of incompatibility in *A Critical Exposition of the Philosophy of Leibniz*.

Despite this endorsement of Kant's account of the status of mathematics, however, Russell is briskly dismissive of the Kantian arguments for the necessity of space, which he had discussed sympathetically in *An Essay on the Foundations of Geometry*. He now writes: 'The theory of necessity urged by Kant... appears radically vicious.'[42] Russell's reasons for this claim are similar to those which he had advanced in his 1898 reply to Couturat and reminiscent of Moore's criticisms of *An Essay on the Foundations of Geometry*. In attempting to vindicate the necessity of space by reference to our experiences and beliefs, Russell writes, 'we only push one stage farther back the region of "mere fact", for the constitution of our minds remains still a mere fact'. In the 1898 reply to Couturat, Russell had combined this rejection of the Kantian treatment of necessity with his intuitionist account of necessity. This same account was expressed in *A Critical Exposition of the Philosophy of Leibniz* but it was there combined with the sceptical thought that all truths are necessary. In *The Principles of Mathematics* there is no sign of the intuitionist account of necessity; but Russell here repeats the sceptical claim that necessity adds nothing to truth:

> On the other hand, there seems to be no true proposition of which there is any sense in saying that it might have been false. One might as well say that redness might have been a taste and not a colour. What is true, is true; what is false, is false; and concerning fundamentals there is nothing more to be said.

Somewhat confusingly, Russell then adds that there is some good sense to the account of necessity advanced by Moore in his 1900 paper 'Necessity', according to which: 'A proposition is more or less necessary according as the class of propositions for which it is a premise is greater or smaller. In this sense the propositions of logic have the greatest necessity, and those of geometry have a high degree of necessity.' Russell does not explain here how the reference to a proposition's being a 'premise' is to be understood. But if one looks back to Moore's paper, it is clear that what Moore had in mind was a generalization of the Kantian conception of the a priori, such that one proposition is 'prior' to, or necessary for, another where the former has a role in justifying the latter. Although Moore writes of the priority here as 'logical priority' the basic thought is epistemological, and concerns the role of more or less fundamental beliefs in the justification of other beliefs. Thus propositions of logic have 'the greatest necessity' in that they are involved in reasoning of all kinds, whereas geometry has a lower degree of necessity since it is involved only in the justification of spatial beliefs. If this interpretation is right, it appears odd that Russell should support it, since in his criticisms of Kant, Russell had, in effect, argued that metaphysical questions concerning necessity should be separated from epistemological ('psychological') questions concerning the a priori. But once one takes account of his modal scepticism and applies this to the metaphysical conception of necessity, one can read his support for this Moorean conception of necessity as a way of saying that

[42] For this and the next three quotations: Russell, *The Principles of Mathematics*, p. 454.

if one really does want to persist with talk of necessity, then one should accept that the only legitimate way to do so is, after all, to give it an epistemological significance. This interpretation of Russell's text is supported by the fact that his 1901 paper 'The Notion of Order and Absolute Position in Space and Time' already contains the passage from *The Principles of Mathematics* which I have been discussing, except that just after the final sentence of the passage cited above ('What is true, is true . . . ' etc.), and before the favourable reference to Moore's account of necessity which follows it directly in *The Principles of Mathematics*, in the 1901 paper Russell had written 'Necessity seems to be a psychological rather than a logical notion'[43]—which he omitted from *The Principles of Mathematics.*

One year later Russell includes a brief discussion of necessity in his series of papers on Meinong. Having criticized Meinong's account of implication, he gives a bald statement of his modal scepticism:

> Connected with this point is a second, namely, that it seems impossible to distinguish, among true propositions, some which are necessary from others which are mere facts.[44]

Although he alludes briefly to Moore's conception of necessity as relative priority, he also introduces a new approach which strikes him as attractive—that the distinction between necessary and contingent truths is associated with that between propositions which are true for all times and those whose truth just concerns the present, past or future:

> I cannot help suspecting that the whole feeling of necessity and contingency has been derived from the fact that a sentence containing a verb in the present tense—or indeed, in the past or future, unless with mention of a particular time—changes its meaning continually as the present changes, and thus stands for different propositions at different times, and as a rule sometimes for true ones, sometimes for false ones.[45]

Russell then generalizes this line of thought, so that it does not just concern times:

> And generally, when a proposition contains a term which we instinctively regard as variable, we feel that the proposition is contingent if some values of the variable make the proposition true, others false. For instance, when we say 'the number of this cab has four figures', we feel that it might have had five, because of all the other cabs we might have taken.[46]

A few pages later he summarizes the whole discussion:

[43] 'The Notion of Order and Absolute Position in Space and Time', translated by G. H. Moore in *The Collected Papers of Bertrand Russell* vol. 3, edited by G. H. Moore (London: Routledge, 1993), 241–58, p. 257.

[44] Russell, 'Meinong's Theory of Complexes and Assumptions', Mind 13 (1904); reprinted in *The Collected Papers of Bertrand Russell* vol. 4, 432–74, p. 435.

[45] Russell, 'Meinong's Theory of Complexes and Assumptions', p. 436.

[46] Russell, 'Meinong's Theory of Complexes and Assumptions', p. 436. Russell's example here is slightly misstated. Instead of writing 'because of all the other cabs we might have taken' he should have written 'because there are other cabs whose numbers have five figures.'

I have already remarked that necessity, in the old modal sense, appears to me inadmissible; there is, however, where propositions containing variables are concerned, a meaning for necessity, and that is, that the said propositions hold for all values of the variable.[47]

Russell is here introducing a new 'meaning for necessity', as the truth of propositions containing variables for 'all values of the variable'; and this is the position that predominates in his later discussions of necessity, such as the passage quoted at the start of this paper from his *Introduction to Mathematical Philosophy*. As I mentioned, a central question to which it gives rise concerns the kinds of value by reference to which it is to be determined whether or not a proposition containing a variable is necessary, and, as we shall see, it is sometimes difficult to determine from Russell's discussion what the answer to this question is to be. But in the context of this first appearance of the position the answer to this question is straightforward. At the start of the series of papers on Meinong, Russell writes that 'the object of a thought, even when this object does not exist, has a Being which is in no way dependent on its being an object of thought'[48] and it is then objects of this kind which are to be values for the variables. This point can, I think, be elucidated by reference to the discussion of 'formal implication' in *The Principles of Mathematics*, even though Russell does not here connect this with the meaning for necessity. Russell says here that a 'formal implication', such as the propositional function: $x$ is a man implies $x$ is mortal, 'is asserted strictly of all possible terms'; and he goes on to explain what terms are in the following passage:

Whatever may be an object of thought, or may occur in any true or false proposition, or can be counted as *one*, I call a *term*. This, then, is the widest word in the philosophical vocabulary. I shall use as synonymous with it the words unit, individual, and entity. The first two emphasize the fact that every term is *one*, while the third is derived from the fact that every term has being, i.e. *is* in some sense. A man, a moment, a number, a class, a relation, a chimaera, or anything else that can be mentioned, is sure to be a term; and to deny that such and such a thing is a term must always be false.[49]

Thus although 'every term has being', Russell does not say that every term exists. Indeed, as in the Meinong paper, he specifically denies this and maintains that there is an 'essential' distinction between existence and being:

Yet this distinction is essential, if we are ever to deny the existence of anything. For what does not exist must be something, or it would be meaningless to deny its existence; and hence we need the concept of being, as that which belongs even to the non-existent.[50]

The conclusion to be drawn, therefore, is that where a propositional function is true for all values of its variable, it is not simply true for values which exist; it is also true for values which do not exist, for all 'possible terms', including the Homeric Gods,

[47] Russell, 'Meinong's Theory of Complexes and Assumptions', p. 450.

[48] In a footnote to this passage Russell writes that he has been led to accept this thesis by Moore; 'Meinong's Theory of Complexes and Assumptions', p. 432, n.2.

[49] Russell, *The Principles of Mathematics*, p. 43.

[50] Russell, *The Principles of Mathematics*, p. 450.

chimaeras, and four-dimensional spaces. In which case it does seem reasonable to regard such a propositional function, or the corresponding universally quantified proposition, as necessary.

Russell's remarks about necessity in the Meinong papers show how his thoughts on this subject were changing during 1903, when the papers were written, though they were not published until 1904. But a brief note written and published in 1904 shows that Russell's thoughts were still changing. This is a note in which Russell responds to some comments on *An Essay on the Foundations of Geometry* by Hugh MacColl which had appeared in *The Athenaeum* the previous year. One comment concerned the definition of possibility, and while accepting MacColl's criticism of the position he had taken earlier, Russell now combines his familiar scepticism concerning necessity and possibility as properties of propositions with a new acknowledgement of an epistemic conception of possibility:

> I should myself maintain that, in an ultimate logical sense—i.e. when all reference to our ignorance is excluded—all propositions are merely true or false. I should not now divide true propositions into necessary and contingent, or false propositions into impossible and possible. Thus, in regard to actual space, I should say that, whether it is Euclidean or whether it is non-Euclidean, there is no sense in saying that it might have been different, but that we do not as yet know which alternative is the true one, and that, in this sense only, either alternative is possible.[51]

## 6.2 'Necessity and Possibility'

As we have seen, in his early writings Russell's views about necessity evolve through (i) the Kantian position of *An Essay on the Foundations of Geometry*; (ii) the intuitionist position of his reply to Couturat; (iii) the modal scepticism in *The Principles of Mathematics* to the effect that the necessity of a proposition adds nothing to its truth; (iv) the suggestion in his comments on Meinong that necessity and possibility are to be thought of instead as properties of propositional functions which are true for all or some values; (v) the recognition that there is also an epistemic conception of possibility, as what is not ruled out by what is known, to which he alludes in his reply to MacColl. In the light of these changes in his own views, it is not surprising that Russell should have attempted a critical assessment of the main positions on the topic of necessity and possibility, which he undertook in the paper 'Necessity and Possibility' that he read to the Oxford Philosophical Society on 22 October 1905. There is evidence that Russell's paper was a revision of an earlier draft from May 1905;[52] but he never revised this paper for publication or even used

[51] Russell, 'Non-Euclidean Geometry', *The Athenaeum* no. 4018 (1904), 592–3; reprinted in *The Collected Papers of Bertrand Russell* vol. 4, 482–5, p. 482.

[52] Russell's 'Necessity and Possibility' was published for the first time in *The Collected Papers of Bertrand Russell* vol. 4, 508–20, and on the issue of the earlier draft, see the editor's comment, p. 507.

parts of it elsewhere—which is surprising but perhaps just indicates the pressure he was under for the next few years to concentrate on mathematical logic.

There is, however, no sign of uncertainty in the way in which Russell opens the paper. He begins by boldly asserting that 'the division of judgments into necessary, assertorical and problematic is in the main based upon error and confusion'.[53] He goes on to say that although there are valid distinctions which correspond to the traditional ones, 'none of them are fundamental, and all of them are better described in non-modal terms'.[54] This sets up a critical agenda, which he introduces by reference to debates about the definition of *good* of the kind discussed by Moore in *Principia Ethica*.[55]

> Different definitions of necessity have been given by different authors; but as a rule these definitions have not been purely verbal. That is to say, the authors have believed that they have an idea of necessity, and that the definitions they gave were *true*, *i.e.* gave marks, other than necessity, which are common and peculiar to what is necessary. If this had not been the case, different definitions would not have been, as in fact they have been, marks of philosophical disagreement. For example, when one writer says that the good is a pleasure, and another that it is virtue, they differ in opinion, because both attach the same meaning to the word *good*, though they differ as to the things that are good.[56]

So far the implication would seem to be that the intuitionist account of necessity he had advanced in some of his early writings is basically correct. But, having just asserted of modal concepts that 'none of them are fundamental, and all of them are better described in non-modal terms', he moves on to the thesis he wants to argue for:

> The main question to be considered in regard to necessity is, therefore: Is there any such predicate as *necessity*, as distinct from the various predicates which various definitions assert to be equivalent to it? If not, different definitions do not disagree philosophically, but only as regards the use of words. I do not myself believe that there is such a predicate as *necessity*, apart from definitions which are strictly verbal definitions; though I hardly see how my opinion is to be *proved*.[57]

So Russell distinguishes *necessity*, as he conceives it, from *good* as Moore conceived it. In the latter case, according to Moore (and Russell at this time) there is a fundamental indefinable property, goodness, which makes it possible for there to be genuine debates concerning the definition of 'the good', that is, concerning the 'marks, other than [goodness], which are common and peculiar to what is [good]'. By contrast, despite the way the passage starts, Russell wants to persuade us to share his modal scepticism, to the effect that there is no such fundamental property of necessity, so that 'different definitions do not disagree philosophically, but only as regards the use of words'.

[53] Russell, 'Necessity and Possibility', p. 508. Russell is here alluding to Kant's distinction in *The Critique of Pure Reason* (A70/B95) between the three 'moments' in the modality of judgement—the 'apodictic' (i.e. necessary), the assertoric (i.e. contingently true), and the problematic (i.e. possible).

[54] Russell, 'Necessity and Possibility', p. 508.

[55] See G. E. Moore, *Principia Ethica* (Cambridge: Cambridge University Press, 1903), particularly §13.

[56] Russell, 'Necessity and Possibility', p. 509.

[57] Russell, 'Necessity and Possibility', p. 509.

To support this thesis Russell turns to a critical discussion of a succession of putative definitions of necessity, his aim being to explore 'what people really have in their minds when they affirm necessity',[58] their intuitive '*feeling* of necessity'[59] that attaches to some propositions and not to others. The first position he considers defines necessity in terms of the a priori. Concerning this definition he comments that whatever the epistemological merits of the a priori/empirical distinction, it does not give rise to a distinction among propositions which captures the feeling of necessity. Russell then turns to the suggestion, which he attributes to Bradley, that what is definitive of necessary propositions is that they are 'demonstrable'. Russell interprets demonstrability in terms of implication and objects to Bradley's definition that, since, for any proposition, there will be some proposition which implies it, this definition makes every proposition, true or false, necessary. And even if it is added that the premises of the demonstration must be true, so that the claim is that *p* is necessary just where 'there is a true proposition which implies *p*', the definition still gives necessity a 'logical meaning [which] is wholly destitute of importance, since it holds when and only when *p* is true'.[60]

Russell does not explain here that he takes implication to be material implication, and as a result his discussion of Bradley is somewhat unfair, since by demonstrability Bradley clearly has in mind the possibility of a demonstration, or proof.[61] But Russell in effect explores Bradley's actual position when he moves on to discuss the next definition, according to which what is characteristic of necessary truths is that they are 'analytic'. Russell is more sympathetic to this position:

> This view of necessity and possibility requires much modification before it can be harmonised with modern logic; but by the help of such modification it can, I believe, be still rendered more or less serviceable.[62]

The most important modification, Russell thinks, concerns the need to distinguish between '*p* implies *q*' and '*q* is deducible from *p*'. Russell here makes it explicit that he takes implication to be material implication, and he distinguishes this from deducibility which he defines as provability from the 'laws of deduction'. Since he formulates these laws of deduction in terms of implication, implication then enters into his account of deducibility: '*q* is *deducible* from *p* if it can be shown *by means of the above principles* [the "*laws of deduction*"] that *p* implies *q*.'[63] It is then in terms of deducibility, thus understood, that he interprets the suggestion that what is distinctive of necessary truth is the fact that it is 'analytic', in that it can be deduced from the fundamental laws of logic.

Before introducing Russell's response to this suggestion, it is worth stepping aside for a moment to compare his remarks here about analyticity with the earlier discussions in *A Critical Exposition of the Philosophy of Leibniz* and *The Principles*

[58] Russell, 'Necessity and Possibility', p. 509.

[59] Russell, 'Necessity and Possibility', p. 510.

[60] Russell, 'Necessity and Possibility', p. 512.

[61] Thus Bradley (*The Principles of Logic*, Oxford: Clarendon Press, 2nd edition, 1922, p. 200) writes: 'Necessity is here the force which compels us to go to a conclusion, if we start from premises.'

[62] Russell, 'Necessity and Possibility', p. 514.

[63] Russell, 'Necessity and Possibility', p. 515.

*of Mathematics* where he had denied that logic is analytic. He begins by observing here that if, by analyticity, one just means deducibility from the law of contradiction (or the laws of contradiction, excluded middle, and identity), then indeed logic is not analytic. But, he now adds, once one recognizes that logic requires more than these laws, it is sensible to modify the meaning of 'analytic' to include all that is deducible from the laws of logic, properly understood; and this definition, he remarks, 'is conformable in the spirit, though not in the letter, to the pre-Kantian usage'.[64] So, he continues, reversing the position he had taken in *The Principles of Mathematics*:

> Certainly Kant, in urging that pure mathematics consists of synthetic propositions, was urging, among other things, that pure mathematics cannot be deduced from the laws of logic alone. In this we now know that he was wrong and Leibniz was right; to call pure mathematics analytic is therefore an appropriate way of marking dissent from Kant on this point.

Russell refers to Couturat as the source of the modified definition of analyticity he adopts here; but it is hard not to see also the influence of Frege, with whose work Russell was by now fully familiar, although there is no reference to it in this paper.

Returning now to Russell's discussion of the suggestion that the mark of a necessary proposition is that it is analytic, Russell argues that this suggestion fails to capture 'the *feeling* of necessity' in a satisfactory way. For, he observes, 'many propositions are felt to be necessary which are not analytic':

> Such are: 'If a thing is good, it is not bad', 'If a thing is yellow, it is not red', and so on. *Bad* does not mean the same as *not-good*, and therefore mere logic will never prove that *good* and *bad* are any more incompatible than *round* and *blue*. Hence, although the class of analytic propositions is an important class, it does not seem to be the same as the class of necessary propositions.[65]

The reference here to the incompatibility of *good* and *bad* is reminiscent of the discussion in *A Critical Exposition of the Philosophy of Leibniz* of the incompatibility of *round* and *square*. So Russell's complaint is that although the modified conception of analyticity ensures that the abstract truths of logic and mathematics count as analytic, and thus necessary, there remain many other propositions which are intuitively 'felt to be necessary' but which are not analytic.

Finally, Russell turns to the suggestion which he had advanced in his Meinong papers that necessity be defined as the mark of propositional functions which hold of everything, and possibility as the mark of propositional functions which hold of something. He now associates this proposal with the position advanced by Hugh MacColl some years earlier,[66] that statements which are necessary truths are always true, and so on, though he comments that MacColl did not make the appropriate distinction between complete statements, i.e. propositions, and propositional functions; the former are just true or false,

[64] Russell, 'Necessity and Possibility', p. 516.

[65] Russell, 'Necessity and Possibility', p. 517.

[66] Hugh MacColl, 'Calculus of Equivalent Statements', *Proceedings of the London Mathematical Society*, vol. 28 (1896), 156–83.

whereas it is the latter which are always, sometimes or never true. Russell's attitude to this suggestion is not altogether clear. Having explained what the position is, he writes:

> There is, so far as I can see, no particular objection to these definitions, except that they do not make necessity and possibility a property of propositions.[67]

The relevance of this remark is that the general position he is attacking is one which takes it that necessity and possibility are primarily properties of propositions; so the position does not provide a definition of necessity of the kind that he has been considering. That fact, however, should be no objection to the position from Russell's point of view. But he does now turn to consider whether the position does justice to our 'feeling of necessity', and introduces a significant objection to it:

> For example, we feel certain of the truth of all propositions of the type: '*x* either is not a moment of time, or is a moment of time subsequent to the death of Cromwell, or is a moment of time preceding the Restoration'; yet we should hesitate to call propositions of this type *necessary*. For we realize at once that the truth of all propositions of this type is a deduction from 'the death of Cromwell preceded the Restoration', which must be a contingent proposition if any proposition is to be contingent.[68]

This certainly looks to be a compelling counterexample to the proposal under consideration. Russell nonetheless suggests that it can be set aside:

> Yet perhaps this feeling could be turned into its opposite. For if anybody said 'such and such an event happened before the death of Cromwell but after the Restoration', we should reply 'that is *impossible*, because Cromwell died before the Restoration'. Thus the feeling of necessity on such points seems to be uncertain and vacillating.

This response is odd. The 'reply' Russell introduces to defend his proposal confuses the necessity of a conditional with the necessity of its consequent. The fact that Cromwell died before the Restoration necessarily implies that no event happened before the death of Cromwell but after the Restoration; nonetheless, since the Restoration might have occurred before Cromwell died, it might have been the case that plenty of events happened before the death of Cromwell but after the Restoration. The occurrence of such events was not unconditionally impossible. But Russell's proposal has precisely this implication: for it interprets 'no event happened before the death of Cromwell but after the Restoration' as 'it is impossible that any event happened before the death of Cromwell but after the Restoration'. So an unconditional impossibility is inferred from a contingent truth and it is this inference which is correctly identified as unwarranted in the objection.

Russell's objection to his own proposal is, therefore, more substantial than he appears to allow. He might reply that in his response to the objection his aim was

[67] Russell, 'Necessity and Possibility', p. 518.

[68] This passage and the following one: Russell, 'Necessity and Possibility', p. 519.

not to defend the proposal but just to show how our intuitive 'feeling of necessity on such points seems to be uncertain and vacillating'. But while it is indeed true that we are prone to confuse the necessity of a conditional with the necessity of its consequent, this is scarcely a good reason for thinking that we should not correct this confusion. Still, Russell's conclusion is that we cannot place any weight on our intuitive feeling of necessity:

> But if any conclusion is warranted by the above arguments, it is this: That the feeling of necessity is a complex and rather muddled feeling, compounded of such elements as the following:
>
> (1) The feeling that a proposition can be known without an appeal to perception;
> (2) The feeling that a proposition can be proved;
> (3) The feeling that a proposition can be deduced from the laws of logic;
> (4) The feeling that a proposition holds not only of its actual subject, but of all subjects more or less resembling its actual subject, or, as an extreme case, of all subjects absolutely.
>
> Any one of these four may be used to found a theory of necessity.... I conclude that, so far as appears, there is no one fundamental logical notion of *necessity*, nor consequently of *possibility*. If this conclusion is valid, the subject of modality ought to be banished from logic, since propositions are simply true or false, and there is no such comparative and superlative of truth as is implied by the notions of contingency and necessity.[69]

At this point it is worth returning to the initial comparison between necessity and goodness. Russell has argued that the four ways of trying to 'define' necessity he has considered are unsatisfactory in one way or another, except perhaps the fourth, and while he does not suggest that he has an 'open question' argument comparable to Moore's argument against naturalistic definitions of goodness to show that no satisfactory definition of necessity can be found, the obvious conclusion to draw would appear to be that necessity (or possibility) is indefinable, as he himself had held only a few years earlier. But instead of this conclusion, Russell concludes that 'the feeling of necessity is a complex and rather muddled feeling', so that 'so far as appears, there is no one fundamental logical notion of *necessity*, nor consequently of *possibility*'.

Why does Russell opt for this sceptical conclusion? His thought seems to be that because our feeling of necessity has been shown to be rather muddled, there is no well-defined notion of necessity. In fact, however, his discussion only showed that the definitions of necessity that he considered did not succeed in capturing all our intuitive judgements of necessity. But the way Russell expresses his conclusion is revealing, as the rejection of the view that the notions of contingency and necessity constitute the 'comparative and superlative of truth'. While it is not obvious how Russell's discussion vindicates this conclusion, there is a different way of arguing for it, which is that this approach to modal notions does not provide a satisfactory way of

[69] Russell, 'Necessity and Possibility', p. 520.

thinking about possibility, since it is not a half-way house to truth or to falsehood nor a third truth-value indeterminate between them. Russell might respond that one of the main advantages of his fourth approach, which treats modal notions as properties of propositional functions, and not propositions, is that it can accommodate possibility properly, in that it holds that a propositional function is possible where it is true of something. So there is good reason to reject a core assumption of the position he has been discussing, that necessity is a property of propositions. Yet while this is indeed a fair point, Russell's own discussion of the fourth approach shows that it does not suffice to capture our intuitive feeling of possibility. For in explaining it, Russell repeats the example he had used in his earlier paper:

> Suppose I take a cab, and its number has five figures: I shall feel that it might have had four figures. In this case, all that is meant seems to be: 'This is a London cab, and some London cabs have numbers consisting of four figures'.[70]

Yet, as before, the obvious objection to this is that even if, in fact, no London cabs have four figure numbers (now or at any time), it still might have been the case that Russell's cab did have such a number. As with the conviction that the Restoration might have occurred before Cromwell's death, our intuitive feeling of possibility concerning these things, although not unchallengeable, is pretty robust. While Russell might say, so much the worse for the modal notion of possibility, even as a property of propositional functions, this looks to be just a dogmatic insistence that it must be possible to capture modal notions in non-modal terms. Although one can understand Russell's route to his sceptical conclusion, it is difficult to be satisfied by it.

One question to consider here is whether there is any connection between Russell's modal scepticism and the position he had just developed concerning Meinongian 'non-entities' in the context of his theory of descriptions. 'On Denoting' was published in October 1905, just before he gave his paper 'Necessity and Possibility' in Oxford, and in this paper Russell famously concludes that:

> The whole realm of non-entities, such as 'the round square', 'the even prime other than 2', 'Apollo', 'Hamlet' etc. can now be satisfactorily dealt with. All these are denoting phrases which do not denote anything.[71]

So the hypothesis to consider is that just as Russell's theory of descriptions enabled him to affirm the non-existence of non-entities without attributing 'being' of any kind to them (contrary to the position he had held in *The Principles of Mathematics*), Russell may have thought it desirable to set aside the traditional conceptions of necessity and possibility as properties of propositions, so that he could dispense with the need to attribute being to non-actual possibilia, while providing an account of

[70] Russell, 'Necessity and Possibility', p. 518. This is the corrected version of the garbled example provided earlier in the Meinong paper; cf. n.46 this chapter.

[71] 'On Denoting' in *The Collected Papers of Bertrand Russell* vol. 4, 415–27, p. 425.

necessity and possibility as properties of propositional functions, which, he believed, would enable him to capture the use of modal idioms in non-modal terms. There is no suggestion to this effect in 'Necessity and Possibility' itself, nor, so far as I am aware, in any other writings from this time; but, as we shall see, his later writings do include some suggestions of this kind. However, it has also to be recognized that the ontological parsimony of his theory of descriptions creates a problem for Russell's new account of modality. I argued earlier that this account, which Russell introduced in his Meinong papers, looked to be prima facie tenable because Russell at that time was content to attribute being to objects of all kinds, including objects which do not exist, so that where necessity is understood as the truth of a propositional functional for all values of its variable, this is to be understood as including all possible instances of the function as well as all actual, existing, ones. But once this ontological generosity is replaced by the parsimony of 'On Denoting', which restricts being to actual existence (assuming that merely possible existents are among the non-entities whose being is excluded by Russell), the challenge to Russell of saving the appearances for his non-modal definition of necessity in terms of the universal truth of the relevant propositional function is acute. An important question for us, therefore, is whether Russell provides a solution to this problem in his later writings.

## 6.3 Later Writings

For the next few years Russell published very little about modality. But in 1910 he wrote a reply to some comments by Bradley on *The Principles of Mathematics* in 'On Appearance, Error and Contradiction'.[72] Among other things he alluded to his disagreement with Bradley concerning relations, and commented:

> This opinion seems to rest upon some law of sufficient reason, some desire to show that every truth is 'necessary'. I am inclined to think that a large part of my disagreement with Mr. Bradley turns on a disagreement as to the notion of 'necessity'. I do not myself admit necessity and possibility as fundamental notions: it appears to me that fundamentally truths are merely true in fact, and that the search for a 'sufficient reason' is mistaken. I can see many ways of defining necessity which will account for its common uses: we may call a proposition necessary when it follows from a proposition known to be true, or when it can be known without empirical evidence, or when what is affirmed would be equally true of any other subject.[73]

This is a much abbreviated version of the position advocated in the 1905 paper 'Necessity and Possibility' discussed just now. As such it indicates that Russell's position had not changed much during this period. But the next passage raises an

[72] F. H. Bradley, 'On Appearance, Error and Contradiction', *Mind* 19 (1910): 153–85; reprinted in Bradley, *Essays on Truth and Reality* (Oxford: Clarendon Press, 1914), 245–73.

[73] Russell, 'Some Explanations in Reply to Mr. Bradley', *Mind* 19 (1910): 373–8; reprinted in *The Collected Papers of Bertrand Russell* vol. 6, edited by J. G. Slater (London: Routledge, 1992), 353–8, pp. 354–5.

important question about it. In *The Problems of Philosophy* (1912), when discussing the status of elementary truths of arithmetic, Russell writes:

> We do not, in fact, feel our certainty that two and two are four increased by fresh instances, because as soon as we have seen the truth of this proposition, our certainty becomes so great as to be incapable of growing greater. Moreover, we feel some quality of *necessity* about the proposition 'two and two are four', which is absent from even the best attested empirical generalisations. Such generalisations always remain mere facts: we feel that there might be a world in which they were false, though in the actual world they happen to be true. In any possible world, on the contrary, we feel that two and two would be four: this is not a mere fact, but a necessity to which everything actual and possible must conform.[74]

Russell here uses Leibnizian idioms to express the thesis that the proposition 'two and two are four' is a necessary truth. It is striking that Russell does not manifest any hesitation concerning the attribution of the 'quality of necessity' to this *proposition*. But since he treats this proposition as if it were a general truth comparable to 'empirical generalisations', we can take him to regard it as a universally quantified propositional function. But what is the domain of the quantifier? The fact that Russell's idioms are Leibnizian suggests that the domain is 'possible worlds'. It is not made clear, however, either in this passage, or elsewhere in the book, what significance is to be attached to this idiom. If we think back to his early work on Leibniz, the position he took there was that possible worlds are totalities of compossible substances, where compossibility is to be understood as the absence of incompatibility, and thus ultimately by reference to the 'ultimate and indefinable' property of necessity. But, as we have also seen, Russell had abandoned this conception of necessity soon afterwards, and the comment in his reply to Bradley, that 'I do not myself admit necessity and possibility as fundamental notions' indicates that he had not changed his mind on this point. Nonetheless, this passage from *The Problems of Philosophy* shows that Russell did accept that there is an important contrast between the actual and the possible. So the question arises as to how he thought this contrast is to be substantiated.

One might comment that because this passage comes from *The Problems of Philosophy*, which is explicitly intended to be an introductory book for the general reader, one should not attach too much significance to it. But this passage is not isolated in Russell's writings of this period. In his 1914 Lowell lectures on *Our Knowledge of the External World*, at the start of lecture VII on 'The Positive Theory of Infinity', in the course of a discussion of the similarity between philosophy and mathematics and their difference from history and natural science, Russell writes:

> Between philosophy and pure mathematics there is a certain affinity, in the fact that both are general and *a priori*. Neither of them asserts propositions which, like those of history and geography, depend upon the actual concrete facts being just what they are. We may illustrate

[74] Russell, *The Problems of Philosophy* (London: Williams & Norgate, 1912), p. 121.

this characteristic by means of Leibniz's conception of many *possible* worlds, of which only one is *actual*. In all the many possible worlds, philosophy and mathematics will be the same; the differences will only be in respect of those particular facts which are chronicled by the descriptive sciences. Any quality, therefore, by which our actual world is distinguished from other abstractly possible worlds, must be ignored by mathematics and philosophy alike.[75]

Russell here uses the idiom of possible worlds to 'illustrate' the way in which philosophy and pure mathematics, unlike history and geography, do not 'depend upon the actual concrete facts being just what they are'. Again it is not made clear what significance is to be attached to this talk of 'illustrating' the differences between the different sciences, but the passage suggests that Russell regards possible worlds primarily as an abstract structure which can be used to illustrate these differences, and different in kind from the actual world itself. Thus he writes here of 'the actual concrete facts being just what they are', and then of the ways in which 'our actual world is distinguished from other abstractly possible worlds'. So the distinction between the actual world and merely possible, non-actual, worlds is also a distinction between the 'concrete' and the 'abstract'.

Russell uses the same Leibnizian idioms in his *Introduction to Mathematical Philosophy* when discussing the status of the axiom of infinity:

From the fact that the infinite is not self-contradictory, but is also not demonstrable logically, we must conclude that nothing can be known *a priori* as to whether the number of things in the world is finite or infinite. The conclusion is, therefore, to adopt a Leibnizian phraseology, that some of the possible worlds are finite, some infinite, and we have no means of knowing to which of these two kinds our actual world belongs. The axiom of infinity will be true in some possible worlds and false in others; whether it is true or false in this world, we cannot tell.[76]

Russell uses his 'Leibnizian phraseology' here to characterize the status of the axiom of infinity as 'true in some possible worlds and false in others'. And although his main conclusion here is the epistemological thesis that 'nothing can be known *a priori* as to whether the number of things in the world is finite or infinite', he introduces modal idioms to explicate the significance of this thesis, as indicating 'that some of the possible worlds are finite, some infinite, and we have no means of knowing to which of these two kinds our actual world belongs'. So this passage implies that these modal idioms, with their contrast between the actual and possible, have some substantive significance in virtue of which they support his sceptical epistemological thesis.[77]

The fact that the passage I used right at the start of this paper, in which Russell combines a denunciation of the 'traditional view' that 'among true propositions,

[75] Russell, *Our Knowledge of the External World* (London: George Allen & Unwin, 1914), p. 186.

[76] Russell, *Introduction to Mathematical Philosophy* (London: George Allen & Unwin, 1919), p. 141.

[77] At the end of the book Russell again writes that 'Among "possible" worlds, in the Leibnizian sense, there will be worlds having one, two, three,... individuals' (*Introduction to Mathematical Philosophy*, p. 203). The use of scare quotes here is, presumably, to indicate that the metaphysical status of possible worlds is not straightforward, though no account of this status is offered.

some were necessary, while others were merely contingent' with his positive affirmation of the view that modal notions can be attributed to propositional functions, comes from his *Introduction to Mathematical Philosophy*,[78] where as we see above, he also employs his 'Leibnizian phraseology', suggests that Russell believed that these aspects of his treatment of modal notions could be combined; but he does not here explain how this is to be done. The obvious way to begin to reconcile these different aspects of his position is to take it that, in some sense, possible worlds, or possibilities of some kind, enter into the domain of the quantifiers expressing the modal notions that are attributed to propositional functions. But what is challenging is to work out how this suggestion should be developed. We saw, for example, in the passage from the Lowell lectures, that Russell distinguishes between the 'concrete facts' of the actual world and 'other abstractly possible worlds'.[79] So it would certainly be a mistake to attribute to him a Lewis-style modal realism.

This point is confirmed by a passage from Russell's 1913 manuscript 'Theory of Knowledge'. In the course of discussing William James's account of verification, which makes reference to possible as well as actual verification, Russell writes:

> It may be laid down generally that *possibility* always marks insufficient analysis: when analysis is completed, only the *actual* can be relevant, for the simple reason that there is only the actual, and that the merely possible is nothing.[80]

This emphatic claim is repeated later when, in the course of discussing the view that just as true propositions assert the existence of real objects, false propositions assert the existence of unreal ones, Russell asserts that 'there cannot possibly be such things as unreal objects', and he then continues:

> It is only necessary to observe that what has been said about unreality applies unchanged when it is called by some title of politeness, such as 'being for me' or 'being for thought', which represent merely the vacillating regret in pronouncing the sentence of non-existence on life-long friends. And the same applies to any philosophy which believes, in any ultimate way, in a realm of 'possibles' which are not actual. The view that the possible is something, but not quite so much something as the actual, and that error consists in mistaking the possible for the actual, is only rendered possible by the wrong analysis of sentences which results from confusing descriptions with proper names.[81]

Thus Russell unequivocally rejects here 'any philosophy which believes, in any ultimate way, in a realm of 'possibles' which are not actual' for the reasons 'that there is only the actual, and that the merely possible is nothing'. Although the content of these passages is not, so far as I am aware, expressed in a similarly unequivocal way

[78] Russell, *Introduction to Mathematical Philosophy*, p. 165.

[79] Russell, *Our Knowledge of the External World*, p. 186.

[80] Russell, 'Theory of Knowledge' in *The Collected Papers of Bertrand Russell* vol. 7, edited by E. R. Eames (London: Routledge, 1984), p. 27.

[81] Russell, 'Theory of Knowledge', p. 152.

in any of Russell's published writings, I think one should accept that the attitude expressed here is a core component of Russell's attitude to modality. In his *Introduction to Mathematical Philosophy* he concludes his dismissive discussion of Meinong by writing that a 'robust sense of reality is very necessary in framing a correct analysis of propositions about unicorns, golden mountains, round squares and other pseudo-objects'.[82] Russell does not explicitly refer here to possible objects as such among these 'pseudo-objects', but it seems reasonable to count golden mountains and unicorns (as well as a present King of France) as possible objects. The 'robust sense of reality' that he proclaims here must surely be taken to exclude 'any philosophy which believes, in any ultimate way, in a realm of "possibles" which are not actual'.[83]

This conclusion leaves open the option of attributing to him the view that, without introducing any non-actual items into one's 'ultimate' domain of quantification, one can construct a way of thinking about possibilities on the basis of items from this ultimate domain that is adequate to capture the intended significance of modal notions. In support of this option one can point to the passage from his *Introduction to Mathematical Philosophy* (p. 170) quoted above, where he alludes to the importance of 'a correct analysis of propositions about... pseudo-objects', for in the subsequent discussion he sets out an approach to these propositions in which he invokes his theory of descriptions to make good sense of talk of them. Even more to the point is the final sentence of the second passage quoted above from the 'Theory of Knowledge' manuscript, where Russell writes: 'The view that the possible is something, but not quite so much something as the actual, and that error consists in mistaking the possible for the actual, is only rendered possible by the wrong analysis of sentences which results from confusing descriptions with proper names.' Russell does not explain the significance of this remark, but it is plausible to take it that the comment about names and descriptions presupposes that if talk about possibilities involved naming them, then in such talk one would be treating possibilities as 'ultimate', as genuine 'somethings'. Russell seems to think that because it would follow that, like true belief, belief in such a possibility would be belief in something real, one would need to add that what is merely possible is not quite as real as what is actual so that the mark of error consists in confusing these two.[84] For this suggests that his view is that if we take it that in our talk about possibilities we are not naming them but just 'describing' them, then we can avoid any commitment to treating possibilities as ultimate. This, then, brings us back to the theory of descriptions. It is a familiar point that, despite Russell's inclination to suggest otherwise, this theory by

[82] Russell, *Introduction to Mathematical Philosophy*, p. 170.

[83] Russell, 'Theory of Knowledge', p. 152.

[84] The modal realist will of course reject this suggestion: he holds that the non-actual is just as real as the actual, and that confusion concerning what is actual and non-actual suffices for error without any need to add that they differ in their degree of reality. But that disagreement is not central to the main point here, which concerns the significance of Russell's comment about confusing descriptions with proper names.

itself has no significant ontological implications: whether we treat 'Sir Walter Scott' as a proper name or as a surrogate for a description of him which captures the way in which we think of him, e.g. as 'the author of Waverley', we still end up with a proposition whose truth involves the existence of a person. So, on the face of it, the question whether in our ordinary talk possibilities are being named or described should make no difference to their implied ontological status, to whether or not it is implied that they are 'ultimate' entities. But Russell's theory of logical fictions, as exemplified by his treatment of apparent reference to classes in terms of statements concerning propositional functions[85] and his theory of logical constructions, as propounded in his paper 'The Relation of Sense-Data to Physics'[86] do have significant implications for the status of the things to which these theories are applied. And despite the differences between them Russell has a tendency to treat these three theories—of descriptions, of logical fictions, and of logical constructions—as though they were all much the same,[87] so it is not unreasonable to take it that the remark here about names and descriptions suggests that the way to make sense of talk about what is possible is to use the methods of logical fiction and/or logical construction to analyse this talk in such a way that its use does not commit one to the existence 'in any ultimate way' of a realm of 'possibles' which are not actual.

## 6.4 'The Philosophy of Logical Atomism'

I shall come back to the question as to whether Russell provides any materials for attributing a position of this kind to him. But first I need to examine Russell's discussion of modality in his 1917–18 lectures on 'The Philosophy of Logical Atomism', since it introduces some fresh considerations. In his fifth lecture, on 'General Propositions and Existence' Russell advances the position that we have already encountered:

> One may call a propositional function
> *necessary*, when it is always true;
> *possible*, when it is sometimes true;
> *impossible*, when it is never true.
>
> Much false philosophy has arisen out of confusing propositional functions and propositions.... The case of *necessary, possible, impossible,* is a case in point. In all traditional philosophy there comes a heading of 'modality', which discusses *necessary, possible,* and *impossible* as

[85] See Russell, *Introduction to Mathematical Philosophy*, chapter XVII.

[86] Russell, 'The Relation of Sense-Data to Physics', *Scientia* 16 (1914): 1–27; reprinted in *The Collected Papers of Bertrand Russell* vol. 8, edited by J. G. Slater (London: Routledge, 1986), 5–26.

[87] See, for example, Russell's comments at the end of lecture 6 ('Descriptions and Incomplete Symbols') of his lectures on 'The Philosophy of Logical Atomism' *The Monist* 28 (1918): 495–527; 29 (1918): 32–63, 190–222, 345–80; reprinted in *The Collected Papers of Bertrand Russell* vol. 8, ed. J. G. Slater (London: Routledge, 1986), p. 221.

properties of propositions, whereas in fact they are properties of propositional functions. Propositions are only true or false.[88]

Russell now continues in a way that appears to remove any distinction between the actual and possible:

In such a statement as 'I met a man', you can understand my statement perfectly well without knowing whom I met, and the actual person is not a constituent of the proposition. You are really asserting there that a certain propositional function is sometimes true, namely the propositional function 'I met $x$ and $x$ is human.' There is at least one value of $x$ for which that is true, and that therefore is a possible propositional function.[89]

Russell's comment that 'the actual person' I met is not a 'constituent of the proposition' confirms the normal assumption that the domain of values for $x$ for which the propositional function 'I met $x$ and $x$ is human' is true comprises objects which actually exist. But since the truth of this propositional function for some value of $x$ is said to amount to its being possible, it follows that possibility is existential truth, the actual existence of something which satisfies the relevant propositional function. And Russell appears to accept this implication in the following passage which occurs shortly after that just quoted:

*Existence.* When you take any propositional function and assert of it that it is possible, that it is sometimes true, that gives you the fundamental meaning of 'existence'. You may express it by saying that there is at least one value of $x$ for which the propositional function is true. Take '$x$ is a man', there is at least one value of $x$ for which this is true. That is what one means by saying that 'There are men', or that 'Men exist'. Existence is essentially a property of a propositional function. (p. 204)

This passage implies existence and possibility are the same property. 'Men exist' means that the propositional function '$x$ is a man' has at least one value, is sometimes true; but, according to Russell, this latter condition also means that the propositional function is possible. So does 'Men exist' mean the same as 'Men are possible'? This result is absurd, but on the face of it, it is an implication of what Russell has said. Thus Russell is open to the complaint that, so far from providing an account of modality, he has simply applied modal terms to logical conceptions such as existence which have no modal content.

Fortunately, someone in the audience at Russell's lecture wrote a letter to him raising this point. For he begins his seventh lecture in the following way:

Before I begin today the main subject of my lecture, I should like to make a few remarks in explanation and amplification of what I have said about existence in my two previous lectures.

[88] Russell, 'The Philosophy of Logical Atomism', p. 203.

[89] Russell, 'The Philosophy of Logical Atomism', p. 203. This example—'I met a man'—runs through Russell's writings: see *The Principles of Mathematics* (pp. 53–4) and 'On Denoting' (p. 416). Russell's remark here that 'the actual person is not a constituent of the proposition' comes from these earlier discussions and is not appropriate to the position Russell propounds in these lectures on 'The Philosophy of Logical Atomism' (see especially lecture 1 on 'Facts and Propositions'). Reference to a constituent of a 'fact' would have been better.

This is chiefly on account of a letter I have received from a member of the class, raising many points which, I think, were present in other minds too.

The first point I wish to clear up is this: I did not mean to say that when one says that a thing exists, one means the same as when one says that it is possible. What I meant was, that the fundamental logical idea, the primitive idea, out of which both those are derived is the same. That is not quite the same thing as to say that the statement that a thing exists is the same as the statement that it is possible, which I do not hold.[90]

Before explaining the distinction he now wants to draw between existence and possibility, however, he attempts to clarify his account of possibility, and it is worth quoting the whole passage in which he does so although it is rather long:

I used the word 'possible' in perhaps a somewhat strange sense, because I wanted some word for a fundamental logical idea for which no word exists in ordinary language, and therefore if one is to try to express in ordinary language the idea in question, one has to take some word and make it convey the sense that I was giving to the word 'possible', which is by no means the only sense that it has but is a sense that was convenient for my purpose. We say of a propositional function that it is possible, where there are cases in which it is true. That is not exactly the same thing as what one ordinarily means, for instance, when one says that it is possible that it may rain tomorrow. But what I contend is, that the ordinary uses of 'possible' are derived from this notion by a process. E.g., normally when you say of a proposition that it is possible, you mean something like this: first of all it is implied that you do not know whether it is true or false, and I think it is implied; secondly, that it is one of a class of propositions, some of which are known to be true. When I say, e.g., 'It is possible that it may rain tomorrow'—'It will rain tomorrow' is one of the class of propositions 'It rains at time *t*', where *t* is different times. We mean partly that we do not know whether it will rain or whether it will not, but also that we do know that that is the sort of proposition that is quite apt to be true, that it is a value of a propositional function of which we know some value to be true. Many of the ordinary uses of 'possible' come under this head, I think you will find. That is to say, that if you say of a proposition that it is possible, what you have is this: 'There is in this proposition some constituent, which, if you turn it into a variable, will give you a propositional function that is sometimes true.' You ought not therefore to say of a proposition simply that it is possible, but rather that it is possible in respect of such-and-such a constituent. That would be a more full expression.[91]

There is much in this long passage that merits attention. First, Russell now allows that it is legitimate to predicate possibility of a proposition, such as that it may rain tomorrow. But, he maintains, this use of 'possible' is derived from a 'fundamental logical idea', namely that there are cases in which an associated propositional function such as 'it rains at time *t*' is true, which he has chosen to express by a new use of the word 'possible'. This admission implies that there is no 'hopeless confusion' in the position of traditional philosophers who have regarded modal notions as properties of propositions; instead, according to Russell, their mistake was just that

[90] Russell, 'The Philosophy of Logical Atomism', p. 222.
[91] Russell, 'The Philosophy of Logical Atomism', p. 222.

they took it that these are the fundamental modal properties—whereas in truth these properties of propositions are derived from some fundamental properties of propositional functions which Russell has chosen to describe by new uses of the familiar modal words. Once the debate is set out in this way, however, it follows that Russell cannot dismiss his opponents in the way in which he is inclined to do; if anything, the burden of proof falls upon him—he needs to be able to substantiate his claim that the properties of propositional functions he has identified are indeed fundamental to propositions themselves having the familiar modal properties.

A second point to note here is Russell's recognition that ordinary uses of 'possible' often have an epistemic content—'it is implied that you do not know whether it is true or false'—which had previously surfaced in Russell's brief 1904 response to MacColl.[92] However, Russell also maintains that we can distinguish between these epistemic possibilities and the non-epistemic possibilities of which he writes that '[m]any of the ordinary uses of "possible" come under this head'. It is, then, these latter possibilities with which he is primarily concerned, and of which he tries to provide a better account than he had done in his previous lecture. But here it is hard not to feel that Russell has not thought through what he is saying. The claim that if you say of a proposition that it is possible, what you have is this: 'There is in this proposition some constituent, which, if you turn it into a variable, will give you a propositional function that is sometimes true' implies that the possibility that something is the case consists simply in the truth of an existential generalization derived from it—'a propositional function that is sometimes true'. This has absurd results; for example, assuming that the number two is a 'constituent' of the proposition that the number two is odd,[93] it follows that it is possible that the number two is odd, since if this constituent is 'turned into' a variable one arrives at the propositional function '$x$ is odd' which is of course 'sometimes true' since some numbers are odd. Thus, so far from substantiating his claim that the properties of propositional functions he has identified are fundamental to propositions themselves having the familiar modal properties, Russell's remarks here suggest that his claim is altogether misguided.

Nonetheless, there is a way of modifying Russell's final remark that 'You ought not therefore to say of a proposition simply that it is possible, but rather that it is possible in respect of such-and-such a constituent' in such a way as to make reasonably good sense of it. For if one introduces his Leibnizian phraseology, and then maintains that 'You ought not therefore to say of a proposition simply that it is possible, but rather that it is possible in respect of the actual world', one can then go on to say that the truth of this latter claim derives from the fact that if you turn the reference to the

[92] *The Collected Papers of Bertrand Russell* vol. 4, p. 482.

[93] Russell, of course, would deny that the number two is a constituent of the proposition that the number two is odd; instead the constituents of this proposition are complex higher-order propositional functions. But one could adapt the reasoning to the proposition as thus conceived to reach the same absurd result.

actual world into a variable, you will obtain a propositional function that is sometimes true (i.e. being possible in respect of the actual world consists in being true in some possible world). This suggestion is certainly not implicit in what Russell says. The question that arises, therefore, is whether a suggestion of this kind can be proposed as an account of what Russell 'should have said' had he thought more carefully about the matter.

Before addressing this question, however, I should complete the discussion of Russell's treatment of the apparent implication of his earlier lecture, that existence and possibility are the same property, which had started him off on this restatement of his account of possibility. In the paragraph which follows the passage just quoted above he writes:

When I say, for instance, that 'Lions exist', I do not mean the same as if I said that lions were possible; because when you say 'Lions exist', that means that the propositional function 'x is a lion' is a possible one in the sense that there are lions, while when you say 'Lions are possible' that is a different sort of statement altogether, not meaning that a casual individual animal may be a lion, but rather that a *sort* of animal may be the *sort* that we call 'lions'. If you say 'Unicorns are possible', e.g., you would mean that you do not know any reason why there should not be unicorns, which is quite a different proposition from 'Unicorns exist.' As to what you would mean by saying that unicorns are possible, it would always come down to the same thing as 'It is possible that it may rain tomorrow.' You would mean, the proposition 'There are unicorns' is one of a certain set of propositions some of which are known to be true, and that the description of the unicorn does not contain in it anything that *shows* there could not be such beasts.[94]

The main claim here is that his account of possibility does not imply that 'lions exist' and 'lions are possible' have the same meaning, because while both statements concern lions, they do so in different ways. The first statement means that the propositional function '*x* is a lion' is sometimes true, and therefore 'possible' (in Russell's new sense of this word); whereas in the second statement 'lions' refers to a '*sort* of animal' and the whole statement is one to the effect that this sort of animal is possible in virtue of the fact that some sorts of animal exist. So while the meaning of the second statement depends on that of the first statement, since it is only in virtue of the existence of animals of some sort that lions are possible, they are distinct. In effect Russell is here construing 'Lions are possible' as 'It is possible that lions exist', and then interpreting this statement along the lines of his own theory as 'The proposition that lions exist is possible with respect to lions', i.e. as a statement whose truth consists in the fact that the propositional function '*x* is a sort of animal' is sometimes true (and therefore itself 'possible' in Russell's new sense of this word).

This is an ingenious proposal, but it is not satisfactory. As so far expounded, it implies that the question of whether lions are possible has nothing whatever to do with lions, with what sort of animal they are. The proposal implies that, for any

[94] Russell, 'The Philosophy of Logical Atomism', pp. 222–3.

hypothetical sort of animal *X*, it suffices for the truth of '*X*'s are possible' that some sort of animal exists. This is absurd. Russell, to do him justice, notices that his proposal is going to have this result, and therefore adds at the end of his account of what is required for the truth of 'Unicorns are possible' the extra condition 'that the description of the unicorn does not contain in it anything that *shows* there could not be such beasts'. However, this condition undermines the proposal. What is supposed to be being provided is an account of what is required for unicorns to be possible, i.e. for it to be possible that there are unicorns. If the account now includes the condition that the description of unicorns should not imply that it is impossible that there be unicorns, the account is no doubt true, but it is completely trivial since it now includes the very notion of modality, the very possibility, that was supposed to be being explained.

## 6.5 Constructing Possibilities

As we saw, although the position Russell initially presents in his 1917–18 lectures on 'The Philosophy of Logical Atomism' is one which simply equates possibility with existence, he withdraws from this claim once he is challenged on this point. He then clarifies his strategy by introducing his conception of possibility as an existential quantifier and saying that 'the ordinary uses of "possible" are derived from this notion by a process', which is explained in the following way: 'if you say of a proposition that it is possible, what you have is this: "There is in this proposition some constituent, which, if you turn it into a variable, will give you a propositional function that is sometimes true." You ought not therefore to say of a proposition simply that it is possible, but rather that it is possible in respect of such-and-such a constituent.' As I observed, this is still unsatisfactory as it stands, but if one introduces the Leibnizian terminology which Russell employs elsewhere (both before and afterwards), it can be converted into a defensible position by starting from the claim that 'You ought not therefore to say of a proposition simply that it is possible, but rather that it is possible in respect of the actual world'. For one can then go on to say that the truth of this latter claim derives from the fact that if you convert the reference to the actual world into a variable whose values are possible worlds, you obtain thereby a propositional function that is sometimes true, so that being possible in respect of the actual world consists in being true in some possible world. It has to be acknowledged that in these lectures Russell never suggests this position—indeed he never introduces the 'Leibnizian phraseology' which he uses a year later in *Introduction to Mathematical Philosophy*. The explanation of this feature is, I presume, his thesis from the 'Theory of Knowledge' manuscript, which I discussed earlier, that 'there is only the actual, and that the merely possible is nothing'.[95] For

[95] Russell, 'Theory of Knowledge', p. 27.

although this thesis is not repeated in the lectures, the position advanced in the final lecture—'Excursus into Metaphysics: What There Is'—does not make any allowance for that which is 'merely possible' in its description of the 'ultimate simples, out of which the world is built' which have 'a kind of reality not belonging to anything else'.[96] For although Russell here argues for 'the reality of things we think unreal, such as phantoms and hallucinations',[97] this is because 'considered in themselves' they are just 'ordinary sense-data', and therefore have 'the most complete and absolute and perfect reality that anything can have', so that 'they are part of the ultimate constituents of the world'. The presumption that all such sense-data are actual is not explicit; but the tone of Russell's discussion is such as to make it unavoidable.

Nonetheless, there is another aspect of this final lecture which is more congenial to providing an account of possibilities, namely Russell's emphasis here on his methods of logical fiction and logical construction as ways of preserving the beliefs which are central to common sense and science while restricting the basic assumptions concerning the ultimate simples out of which the world is built. Thus, concerning his conception of numbers as classes of classes, he remarks: 'Therefore you do not have, as part of the ultimate constituents of your world, these queer entities that you are inclined to call numbers.' And, he continues, 'the same applies in many other directions.'[98] Russell does not include beliefs about possibilities among these 'other directions'; but bearing in mind that his 'Theory of Knowledge' thesis that 'the merely possible is nothing' can be interpreted as rejecting only 'any philosophy which believes, in any ultimate way, in a realm of "possibles" which are not actual',[99] it is not out of the question that the methods of logical fiction and construction should be applicable to beliefs about possibilities.

So far as I am aware, however, there is only one place where Russell suggests a position of this kind. This occurs in the 'Theory of Knowledge' manuscript:

> When we were discussing relations, we said that, with a given relation and given terms, two complexes are 'logically possible'. But the notion of what is 'logically possible' is not an ultimate one, and must be reduced to something that is *actual* before our analysis can be complete. Now although we do not yet know what a proposition is, it is obvious that what we had in mind, when we said that a complex was 'logically possible', may be expressed by saying that there is a proposition having the same verbal form. This is still not ultimate, because of our doubt as to how propositions are to be explained; but for present purposes we will treat it as ultimate. Of the two 'logically possible' *complexes* '*A* precedes *B*' and '*B* precedes *A*', at most one can be actual; whereas there are always the two *propositions* expressed by these words, in whatever sense there ever are propositions at all.[100]

[96] Russell, 'The Philosophy of Logical Atomism', p. 234.
[97] Russell, 'The Philosophy of Logical Atomism', p. 238.
[98] Russell, 'The Philosophy of Logical Atomism', p. 234.
[99] Russell, 'Theory of Knowledge', p. 152.
[100] Russell, 'Theory of Knowledge', p. 111.

Given Russell's difficulties in his 'Theory of Knowledge' concerning propositions, the details of this position are somewhat indeterminate, as Russell himself acknowledges. But as he also notes, it suffices here to fix the position if we focus on the 'verbal form' of propositions, since we can obtain thereby actual representations of merely possible 'complexes', i.e. of mere possibilities such as that *B* precedes *A*, where in fact *A* precedes *B*. Thus, to put the point in Lewisian terms, although modal realism offends against Russell's robust sense of reality, Russell here seems to entertain the hypothesis of a form of quasi-linguistic modal ersatzism. Yet what is then deeply frustrating about Russell's writings on this topic is the fact that he never develops and discusses a position of this kind.

One work in which a position of this kind is developed is Wittgenstein's *Tractatus Logico-Philosophicus*.[101] Unlike Russell's lectures on 'The Philosophy of Logical Atomism', Wittgenstein's book is thickly imbued with references to possibility, and he makes it clear that he disagrees with Russell on this subject:

> 5.525 It is incorrect to render the proposition '$(\exists x).fx$' in the words, '*fx is possible*', as Russell does.
>
> The certainty, possibility, or impossibility of a state of affairs is not expressed by a proposition, but by an expression's being a tautology, a proposition with sense [*ein sinnvoller Satz*], or a contradiction.

In order to understand this one should bear in mind that propositions are logical pictures of possible states of affairs (2.201, 4.1). So Wittgenstein's point is that what underlies the possibility of a state of affairs is the fact that the proposition which represents it is a meaningful concatenation of names; it has nothing whatever to do with existential quantification in the way that Russell supposed. Wittgenstein's position can be read as a development of the position sketched by Russell himself in his 'Theory of Knowledge'. What is central to it is the thesis that names of individuals and properties are simple, so that any appropriate concatenation of them is a meaningful atomic proposition which represents a possible states of affairs (3.23). On this basis Wittgenstein then uses his logical theory to show how a possible world can be constructed from totalities of these propositions (4.26). Thus in the *Tractatus* Wittgenstein provides a logical construction of possibilities of a kind which, one might have thought, at least merited some attention from Russell.[102] But in his 'Introduction' to the *Tractatus* Russell does not respond to Wittgenstein's criticism of his treatment of possibility, nor does he say anything about Wittgenstein's alternative theory;[103] and in his subsequent writings he just continues to present his

101 L. Wittgenstein, *Tractatus Logico-Philosophicus*, translated by C. K. Ogden (London: Routledge, 1922).

102 Wittgenstein's position was revived by David Armstrong in *A Combinatorial Theory of Possibility* (Cambridge: Cambridge University Press, 1989). The resulting position was assessed by T. Sider in 'Another Look at Armstrong's Combinatorialism', *Noûs* 39 (2005): 680–96.

103 Russell, 'Introduction' in Wittgenstein, *Tractatus Logico-Philosophicus*, pp. 7–23.

usual account of possibility, the position that Wittgenstein had criticized in the *Tractatus.*[104] He indicates by his use of Leibizian idioms that he recognizes the importance of the distinction between what is actual and what is merely possible, but he never provides a logical construction of possibilities which shows how to get beyond his complaint that '*possibility* always marks insufficient analysis' so that 'when analysis is completed, only the *actual* can be relevant, for the simple reason that there is only the actual, and that the merely possible is nothing'.[105] One can only conclude that on this topic, regrettably, Russell left a fundamental gap in the execution of his 'logical-analytic method of philosophy'.[106]

## Bibliography

Armstrong, D., *A Combinatorial Theory of Possibility* (Cambridge: Cambridge University Press, 1989).

Bradley, F. H., 'On Appearance, Error and Contradiction', *Mind* 19 (1910): 153–85.

Bradley, F. H., *Essays on Truth and Reality* (Oxford: Clarendon Press, 1914).

Bradley, F. H., *The Principles of Logic*, 2nd edition (Oxford: Clarendon Press, 1922).

Couturat, L., 'Review of B. Russell *An Essay on the Foundations of Geometry*', *Revue de métaphysique et de morale* 6 (1898): 354–80.

Frege, G., *The Foundations of Arithmetic*, trans. J. L. Austin (Oxford: Blackwell, 1953).

Griffin, N., *Russell's Idealist Apprenticeship* (Oxford: Clarendon Press, 1991).

Hume, D., *A Treatise of Human Nature*, edited by L. A. Selby-Bigge (Oxford: Clarendon Press, 1888).

Lewis, D. K., *On the Plurality of Worlds* (Oxford: Blackwell, 1986).

Loemker, L., *G. W. Leibniz: Philosophical Papers and Letters*, 2nd edition (Dordrecht: Reidel, 1969).

MacColl, H., 'Calculus of Equivalent Statements', *Proceedings of the London Mathematical Society* 28 (1896): 156–83.

Moore, G. E., 'Critical Notice of B. Russell *An Essay on the Foundations of Geometry*', *Mind* 8 (1899): 397–405.

Moore, G. E., 'The Nature of Judgment', *Mind* 8 (1899): 176–93.

Moore, G. E., 'Necessity', *Mind* 9 (1900): 289–304.

Moore, G. E., *Principia Ethica* (Cambridge: Cambridge University Press, 1903).

Moore, G. E., *Early Philosophical Writings*, edited by T. Baldwin & C. Preti (Cambridge: Cambridge University Press, 2011).

Russell, B., *An Essay on the Foundations of Geometry* (Cambridge: Cambridge University Press, 1897).

Russell, B., 'Les Axiomes propres à Euclide, sont-ils empiriques?', *Revue de métaphysique et de morale* 6 (1898): 759–76. Translated by N. Griffin and G. H. Moore as 'Are Euclid's axioms empirical?', in *The Collected Papers of Bertrand Russell* vol. 2, pp. 325–338.

[104] See, for example, *The Analysis of Matter* (London: Allen & Unwin, 1927), p. 170, and *An Inquiry into Meaning and Truth* (London: Allen & Unwin, 1950), especially pp. 37, 182.

[105] Russell, 'Theory of Knowledge', p. 27.

[106] Russell, *Our Knowledge of the External World*, p. v.

Russell, B., *A Critical Exposition of the Philosophy of Leibniz* (London: Allen & Unwin, 1900).

Russell, B., 'L'idée d'ordre et la position absolue dans l'espace et le temps', *Bibliothèque du Congrès International de Philosophie* 3 (1901): 241–77. Translated by G. H. Moore as 'The Notion of Order and Absolute Position in Space and Time', in *The Collected Papers of Bertrand Russell* vol. 3, 241–58.

Russell, B., *The Principles of Mathematics* (London: Allen & Unwin, 1903).

Russell, B., 'Meinong's Theory of Complexes and Assumptions', *Mind* 13 (1904): 204–19, 336–54, 509–24. Reprinted in *The Collected Papers of Bertrand Russell* vol. 4, 432–74.

Russell, B., 'Non-Euclidean Geometry', *The Athenaeum* no. 4018 (1904): 592–3. Reprinted in *The Collected Papers of Bertrand Russell* vol. 4, 482–5.

Russell, B., 'On Denoting', *Mind* 14 (1905): 479–93. Reprinted in *The Collected Papers of Bertrand Russell* vol. 4, 415–27.

Russell, B., 'Some Explanations in Reply to Mr. Bradley', *Mind* 19 (1910): 373–8. Reprinted in *The Collected Papers of Bertrand Russell* vol. 6, 353–8.

Russell, B., *The Problems of Philosophy* (London: Williams & Norgate, 1912).

Russell, B., *Our Knowledge of the External World* (London: George Allen & Unwin, 1914).

Russell, B., 'The Relation of Sense-Data to Physics', *Scientia* 16 (1914): 1–27. Reprinted in *The Collected Papers of Bertrand Russell* vol. 8, 5–26.

Russell, B., 'The Philosophy of Logical Atomism', *The Monist* 28 (1918): 495–527; 29 (1918): 32–63, 190–222, 345–80. Reprinted in *The Collected Papers of Bertrand Russell* vol. 8.

Russell, B., *Introduction to Mathematical Philosophy* (London: George Allen & Unwin, 1919).

Russell, B., *The Analysis of Matter* (London: Allen & Unwin, 1927).

Russell, B., *An Inquiry into Meaning and Truth* (London: Allen & Unwin, 1950).

Russell, B., *My Philosophical Development* (London: Allen & Unwin, 1959).

Russell, B., *The Collected Papers of Bertrand Russell* vol. 7, edited by E. R. Eames (London: Routledge, 1984).

Russell, B., *The Collected Papers of Bertrand Russell* vol. 8, edited by J. G. Slater (London: Routledge, 1986).

Russell, B., *The Collected Papers of Bertrand Russell* vol. 2, edited by N. Griffin & A. C. Lewis (London, Routledge, 1990).

Russell, B., *The Collected Papers of Bertrand Russell* vol. 6, edited by J. G. Slater (London: Routledge, 1992).

Russell, B., *The Collected Papers of Bertrand Russell* vol. 3, edited by G. H. Moore (London: Routledge, 1993).

Russell, B., *The Collected Papers of Bertrand Russell* vol. 4, edited by A. Urquhart (London: Routledge, 1994).

Sider, T., 'Another Look at Armstrong's Combinatorialism', *Noûs* 39 (2005): 680–96.

Strawson, P., *Individuals* (London: Methuen, 1959).

Wittgenstein, L., *Tractatus Logico-Philosophicus*, translated by C. K. Ogden (London: Routledge, 1922).

# 7

# Modality and Degrees of Truth

## An Austro-Polish Sideline in Twentieth-Century Modal Thought

*Peter Simons*

The standard way of conceiving of alethic modality (necessity, possibility, impossibility, contingency, etc.)—'alethic' meaning that it is concerned with the notions of truth and falsity rather than knowledge or belief or nature or psychology or other modality—understands necessary truths as true in a way which is somehow stronger than merely contingent truths. Necessary truths are true and could not have been false, whereas contingent truths are true but could have been false. Put in this way, it is in general 'harder' to be necessarily true than to be true. This can be expressed generally by the two semantic principles:

[1] Any proposition that is necessarily true is true

[2] Some proposition is true but not necessarily true.

We can transcribe these semantic principles to object-language principles by semantic descent:

[3] For all *p*, if necessarily *p* then *p*

[4] For some *p*, *p* and not necessarily *p*.

Taking possibility to be the dual of necessity, we have:

[5] For all *p*, possibly *p* if and only if not necessarily not *p*.

It follows from [3], [4] and [5] by simple logic that:

[6] For all *p*, if *p* then possibly *p*

[7] For some *p*, not *p*, and possibly *p*.

It is then 'easier' to be possibly true than to be true, and 'harder' to be impossibly true (necessarily false) than to be merely not true (false). Defining contingency as:

[8] For all *p*, it is contingent whether *p* if and only if not necessarily *p* and not necessarily not *p*

which by [3] is equivalent to

[9] For all *p*, it is contingent whether *p* if and only if possibly *p* and possibly not *p*

then according to this view, and assuming the principle of bivalence

[10] Every proposition is either true, or false, and none is both

there are four possible and mutually exclusive cases for propositions with regard to their truth and modal status:

Necessarily true
Contingently true
Contingently false
Necessarily false.

None of the principles stated so far entails that there *are* any necessarily true propositions (we shall come back to this point later), but it is almost universally held that some propositions are necessarily true, and therefore also that some are necessarily false. Assuming, as [2] indeed implies, that some propositions are contingently true, and therefore that some are contingently false, this means that all four statuses are exemplified.

This is, as I said, the standard view, and it goes back to Aristotle. What it does not tell us, among other things, is *why* necessities (necessary truths and necessary falsehoods) are necessary. There have been several theories as to why necessities are necessary.

One is that their truth or falsity turns on something to do with language. The most frequently encountered expression of this is that they are true (or false) 'in virtue of meaning'. If we construe propositions as linguistic items, sentences, or speech items, utterances, this makes some sense, perhaps a different sense for the two construals, but it does make sense to say a sentence has a meaning, and it makes sense to say that an utterance means something. If on the other hand propositions are construed, as they often are, in the vein of Bolzano, Frege and Church, as themselves *being* meanings, then the phrase 'true in virtue of meaning' must be reinterpreted, somewhat along the lines of 'this proposition is true in virtue of being the very proposition that it is', which is unenlightening. The implied contrast of course is with a proposition that is true in virtue *both* of being the proposition that it is *and* the way things are in the world. Without going into details, let us call such accounts, which were notably popular among the logical empiricists, *linguistic* accounts of modality.

On the other hand there are theories which hold that necessary truths are ones which are true in every possible circumstance, or in every possible world. Again, there are numerous ways to interpret this, and whole books have been written about the varieties. In some variants, the two accounts are compatible with one another; in others, it is *because* of a proposition's being true 'in' or 'of' every possible world that it is necessarily true. Call such accounts which trace the source of a proposition's modal status to its truth-status across possible worlds *polycosmic* accounts of necessity.

Leaving aside sceptical or eliminative views of alethic modality, such as we find in Russell and Quine, which deny that modality is a 'respectable' notion, or interpret it in some way which eliminates it, most modern accounts of modality fall into one or other of these two camps: linguistic or polycosmic. But not quite all: in this paper I shall discuss two twentieth-century theories of alethic modality which are notably distinct from either of the two main camps. While differing in several ways, they have enough in common to count as a 'third way' of construing modality. Since they both involve truth values for propositions other than the classic ones of truth and falsity, I shall call them *polyvalent* accounts of modality. One is due to the Austrian philosopher Alexius Meinong (1853–1920), the other to the Polish logician Jan Łukasiewicz (1878–1956). This justifies the term 'Austro-Polish' in our subtitle. But there is more to the linkage than this. Both men were born in the Hapsburg Empire, and indeed, as it happened, in the same city: Lwów (L'viv, Lvov, Lemberg), then in Austrian Galicia, later in the Second Republic of Poland, then in the Soviet Union, and today in Ukraine. Meinong had been a student of Franz Brentano in Vienna. Łukasiewicz had been a student of another Brentano student, Kazimierz Twardowski, in Lwów. Neither died in that polity: Meinong died in Graz in the young Republic of Austria, while Łukasiewicz, the bulk of whose career was spent in inter-war Poland, died in exile in the Republic of Ireland. They also knew one another: Łukasiewicz spent parts of 1908 and 1909 in Graz on a research scholarship, interacting with Meinong and other members of the Graz School, and it is a reasonable conjecture that the partial similarities of their views owe something to their interaction. Having said that, I will first examine their views separately, and only then draw the parallels.

## 7.1 Meinong on Probability…

Meinong is known among philosophers for holding the apparently absurd view that there are objects which do not exist. His writings were sufficiently striking to impress the young Bertrand Russell,[1] who at first found them very congenial if not wholly convincing, but as time went on and Russell developed his own views, in particular his theory of definite descriptions, developed in part expressly against Meinong, he became increasingly critical and even carelessly dismissive of Meinong's views. His verdict influenced generations of philosophers who had never read Meinong, but a closer examination of the exchange shows that it was much less one-sided than generally thought.[2] The main objection that Russell and others voiced is not to the idea of non-existent objects as such, but to the idea of contradictory objects, which

[1] Bertrand Russell, 'Meinong's Theory of Complexes and Assumptions', *Mind* 13 (1904): 204–19, 336–54, 509–24.

[2] See my 'Über das, was es nicht gibt: Die Meinong–Russell Kontroverse', *Zeitschrift für Semiotik* 10 (1988), 399–426. Translation: 'On What There Isn't: The Meinong–Russell Dispute', in P. Simons, *Philosophy and Logic in Central Europe from Bolzano to Tarski* (Dordrecht: Kluwer, 1992), 159–92.

*cannot* exist. In fact Meinong's account of modality, developed by him several years after his exchange with Russell, makes no use of contradictory objects, so we can leave that contentious matter on one side.

Meinong's account of modality has its basis in ontology, or in what he preferred to call the *theory of objects*. An object (*Gegenstand*) is anything whatsoever that can be thought of, no matter what its kind or category, and no matter whether it exists or not. Among objects there are two fundamental categories that are important for our purposes. *Things* (Meinong's word is *Objekte*)[3] are what are presentable by ideas and referred to by names. They include individual material things like planets and rivers, people, animals, properties like whiteness or squareness, and relations like loving or being larger than. *Situations* (Meinong's term is *Objektive*)[4] are what is presentable by judgements or assumptions and expressible by a declarative sentence or a clause. To describe a situation we may use a sentence, as *Jupiter is larger than Saturn*, while to refer to it we may use a gerund, *Jupiter's being larger than Saturn*, or a that-clause, as *that Jupiter is larger than Saturn*.

A situation involves things and other objects, which Meinong calls the situation's *material*: the situation of Jupiter's being larger than Saturn involves three things: two planets and a relation. In keeping with Meinong's view that only some objects exist, some situations exist and others do not. Jupiter's being larger than Saturn exists, whereas Saturn's being larger than Jupiter does not, yet it is still an object, and still involves the same three things. Situations that exist are commonly called *facts*, and Meinong is happy to use the cognate German word *Tatsache* for such situations. Situations that do not exist he called *Untatsachen*, so we call them *unfacts*. Incidentally for Meinong the only difference between a true proposition or truth on the one hand and a fact on the other is that a truth is a fact that is apprehended by someone, whereas a fact (factual situation) need not be. *Mutatis mutandis* this applies also to false propositions and unfacts. Propositions and situations are not distinct kinds of object for Meinong.

Among the things that do not exist, which include merely accidentally non-existent things, like the King of France in 1905, and contradictory things like the largest prime number, are things which are incomplete with respect to some property or other (where under 'property' we also include relational properties or determinations such as being larger than Saturn). Actually existing things are always fully determinate according to Meinong: at a given time Jupiter has an absolutely determinate mass, location, rate of rotation, distance from other bodies and so on. Incomplete objects however are not fully determinate, and so cannot exist. Prominent examples

[3] J. N. Findlay, in *Meinong's Theory of Objects and Values* (Oxford: Clarendon Press, 1963), uses 'objecta' but this is very artificial.

[4] The word 'objective' could also be used in a semi-technical way. Often the term 'state of affairs' is used, though Meinong rejected its German equivalent *Sachverhalt* because he thought that term connoted truth or factuality. Nowadays the connotation is less prevalent. But 'situation' is shorter, and close enough in meaning to serve.

of incomplete objects are creatures of fiction. Hamlet has some properties, for example, he is a Danish prince, he is indecisive, he is a good swordsman; but in many respects he is undetermined: he is neither left-handed nor right-handed, nor ambidextrous, he is neither blond nor brunet nor of any other hair-colour, he has no determinate height, weight, and so on. Incomplete objects are useful to Meinong because they correspond exactly to the characteristics of the acts of thought by which we present them to ourselves. If I am asked to imagine an indecisive Danish prince called 'Hamlet', while I may go a little beyond Shakespeare's text in how I imagine my Hamlet to be, I never fully determine him in all respects. Sometimes however an act corresponds to or 'hits' a real and fully determinate thing, as when I see a real person, such as Crown Prince Frederik of Denmark, though even then the content of my act only corresponds to a few aspects of the real person. Meinong says the incomplete object, which exactly corresponds to the content of my partial perception, is an *auxiliary* object (*Hilfsgegenstand*), which as it were helps me to be in cognitive contact with the real or target object (*Zielgegenstand*). Auxiliary objects have a further role as meanings of general words for Meinong.

Meinong finds another ingenious use for incomplete objects in his account of probability. Suppose I am asked to draw a single card from a complete pack 'at random', as they say. What are the chances that I draw an ace? Or a spade? A court card? We know from elementary combinatorial calculation that the chances are, respectively $\frac{1}{13}$, $\frac{1}{4}$, and $\frac{3}{13}$. But when we do draw a card, it either is a spade, or it is not, it is an ace, or not, it is a court card, or not. The actual draw and outcome are fully determinate. If *this* draw draws a spade, the chances of this draw's being a spade are 1, and if this draw draws a card of another suit, the chances of this draw's being a spade are zero. If all I look at are actual situations, there appears to be no place for probability to get a grip. Standard interpretations of probability deal with this by considering not just actual situations but also a range of possible but not actual outcomes, the whole being called the *sample space*. It is important that Meinong does not do this (although he had the resources to do so had he wanted).

Consider now the incomplete object *my draw of a card from pack P at time t*, it being clear which occasion this is. (For brevity henceforth I shall just say 'my draw'.) My actual draw is a spade, or it is not, but this information is not determined by the incomplete object. So now consider the sentence:

My draw is a spade.

This can be understood in one of two ways, depending on whether we take the subject-expression 'my draw' to refer to my actual (and thus complete) draw, or whether we take it to refer to the incomplete object *my draw*.[5] In the former case the sentence is true or false, depending on what actually happens. In the latter case

[5] Using italics as a nonce way to mark the incomplete object.

however Meinong says the sentence is not true, but not false either. Rather it has an intermediate value, determined by the statistics of cases. Taking truth to be value 1 and falsehood to be value 0, the sentence with the incomplete subject has value 0.25. Probability for Meinong is having some degree of truth between 0 and 1, inclusive. So Meinong allows certain propositions to have truth values other than true and false. In other words, he subscribes to a semantics in which there are more than the classical two values, a many-valued or polyvalent semantics. The degrees of truth or probability correspond to the proportion of favourable cases to all cases, and are thus numbers $m$ in the range $0 \leq m \leq 1$.

How these numbers are determined is an intricate question, on which Meinong is not very clear. But the general picture is something like this: the incomplete object *my draw* is logically contained in (Meinong says *implected in*) more complete objects, such as for example *my draw being the 7 of diamonds*, and each of these is implected in a range of ever more determined objects, until we reach objects which are complete and consistent (not formally or materially contradictory). Suppose I do actually draw one card, and suppose it is the 7 of diamonds. Then one of the complete objects in which *my draw being the 7 of diamonds* is implected (there are many) actually exists, so the incomplete objects implected in the actual draw gain a vicarious toehold on being that Meinong calls *implexive being*. All the other infinitely many possible completions of the object *my draw* do not exist. So it looks as though there is an infinitesimal chance of my drawing the card as I do. But this is nonsense. Nearly all of the detail about the completions of *my draw* is irrelevant to which card it is, bearing on e.g. how exactly it is withdrawn, at what angle, from where in the pack, with which hand, and so on. All that matters is which card it is. So there are 52 families of completions, each relating to a different card being drawn. All this assumes of course that the pack is normal and there are just the 52 usual cards. If there is no bias in how the cards are drawn, it is fair to say there is a 1 in 52 chance of any of the cards being drawn, and as there are 13 ways to draw a different spade, the truth value of:

My draw is a spade

when the subject term refers to *my draw* is $\frac{1}{4}$ or 0.25. And this is so whichever card I actually draw: the subject of the proposition with a truth value other than 1 or 0 is not the actual draw but the incomplete object. It is here among incomplete objects, says Meinong, that probability is 'at home'.[6] If the pack is dodgy or there is some other funny business involved, e.g. if all the spades are very slightly larger than the others, or they are presented in a biased way, this will not be so. Meinong's theory is in principle able to cope with non-equiprobable cases, and with many other complications, though he does not go into detail. As Meinong readily admits, despite the

[6] Alexius Meinong, *Über Möglichkeit und Wahrscheinlichkeit. Beiträge zur Gegenstandstheorie und Erkenntnistheorie* (Leipzig: Barth, 1915; reprinted Graz: Akademische Druck- u. Verlagsanstalt, 1972), p. 167.

book in which he expounds his theory being over 700 pages long, it does not go beyond elementary considerations.[7]

## 7.2 . . . or Rather, Possibility

The title of the large book in question is *Über Möglichkeit und Wahrscheinlichkeit* (*On Possibility and Probability*). It was completed in 1913 and published in 1915, and it does indeed deal with probability in what Meinong calls 'the broad sense'. But Meinong is sensitive to the distinction between subjective probability, degree of belief or conviction, and objective, or as he calls it, *unsubjective* probability. He prefers to reserve the term *Wahrscheinlichkeit*, true-seemingness, for subjective probability, and employs *Möglichkeit*, 'possibility' for the objective or object-theoretic sense, which is what we have been describing. The book correspondingly falls into two parts, the first dealing with objective possibility, the second with subjective probability, and along with its dual title bears the dual subtitle *Beiträge zur Gegenstandstheorie und Erkenntnistheorie* (*Contributions to the Theory of Objects and Theory of Knowledge*).

So we have to admit that what we described in the previous section as Meinong's theory of probability is not what *he* would have unqualifiedly called 'probability' but what he would have called 'possibility'. The overall motivation is, however, clearly to give an account of what we call probability (Meinong says, 'in the broad sense'), from both its ontological and its (distinct) epistemological side.

The sense of 'possible' in which a situation or proposition about an incomplete object has a degree of truth between 0 and 1 is what Meinong calls *increasable* possibility. There is also an 'unincreasable' sense of 'possible' in which any situation which is neither factual nor unfactual, but somewhere between, a status Meinong calls *subfactual*, is simply possible.[8] By this count then, ignoring degrees of truth or factuality, there are just three factuality statuses (factual, subfactual, unfactual) or truth values (true, possible, false). Notice that Meinong avoids postulating a fourth status of superfactuality, or being both true and false, since his concern is with probability and its application, and also because, talk of impossible objects and the accusations of Russell notwithstanding, he is concerned to provide a propositionally consistent theory. We might also want the slightly weaker and more extensive use of 'possible' whereby what is true also counts as possible. Meinong calls the possibility of a truth *Auchmöglichkeit*, 'also-possibility' as distinct from the possibility of a non-truth, which is a *Nurmöglichkeit*, 'only-possibility'. Let's make an artificial use of the distinction between upper and lower case and say then that:

[11] A proposition is Possible iff it is neither true nor false

[12] A proposition is possible iff it is true or Possible (i.e. iff it is not false).

[7] Meinong, *Über Möglichkeit und Wahrscheinlichkeit*, p. 728.
[8] Meinong, *Über Möglichkeit und Wahrscheinlichkeit*, p. 73.

In what Meinong calls the *factuality line* and the *possibility line*,[9] the upper limits are, respectively, factuality and truth. There is no *Übertatsächlichkeit* or superfactuality (supertruth) meriting the term 'necessity'. Nothing is more true than truth, or, we might dually add, more false than falsehood. This is contrary to the standard or Aristotelian conception of necessity with which we began. From Meinong's point of view, necessity, understood as the dual of possibility, is no stronger than truth, for by [5], a proposition is necessary if and only if its negation is not possible, i.e. its negation is false, i.e. it is true.

Nevertheless, Meinong is prepared to concede that we do have use for a notion of necessity. But he affords it a different status and a different source than that of possibility as discussed to date. Some truths, indeed some falsehoods and some possibilities, are evident as such. Or rather, since Meinong wishes to keep ontological and epistemic status separate, their truth, falsehood or possibility is written into them as part of their own nature. Meinong calls such truths, etc. *inhesive*.[10] They are true, false, etc. because of being the situations they are. Had Meinong distinguished propositions from situations, he might have called them analytic truths, falsehoods, and so on. So necessity is simply inhesive factuality, and impossibility is inhesive unfactuality. There are also inhesive subfactualities or possibilities, e.g. that *my draw* (the incomplete object) is a spade. A non-inhesive situation Meinong calls *adhesive*. If a situation is such that its truth value does not spring from its very nature, for example that my actual next draw is a diamond, then that is an adhesive truth. Adhesive situations are ones which by their nature could have a different status from the one they in fact have, and correspond therefore closely to contingent propositions.

Despite this, Meinong is not prepared to concede that there are truths stronger than other truths. What are sometimes called necessary truths, such as those of mathematics, are simply inhesive, not supertrue, and the ideal objects of mathematics do not exist necessarily, they just exist (albeit not in space and time).

## 7.3 Many-Valued Meinong

Meinong was not a logician, and so did not work out the consequences for logic of admitting a third truth-status alongside true and false, or an ordered continuum of truth-statuses in the case of increasable possibility. His Graz 'Minister for Logic', Ernst Mally, was too busy working on his own alternatives to Meinong's object and probability theory to help him out, so the status of Meinong's theory as a forebear of three-valued logic and fuzzy logic was not obvious to himself or others at the time.[11]

[9] Meinong, *Über Möglichkeit und Wahrscheinlichkeit*, p.147.

[10] Meinong, *Über Möglichkeit und Wahrscheinlichkeit*, p. 143.

[11] I have found no earlier recognition of this than my own 'Łukasiewicz, Meinong, and Many-Valued Logic', in *The Vienna Circle and the Lvov-Warsaw School*, edited by K. Szaniawski (Dordrecht: Kluwer, 1989), 249–91; reprinted in Simons, *Philosophy and Logic in Central Europe from Bolzano to Tarski* (Dordrecht: Kluwer, 1992), 193–225.

Nor is it clear that, apart from some rudimentary considerations of conjunction and disjunction in the second part of his book on probability, that he was aware of the pitfalls lurking in the attempt to treat probability from the perspective of a polyvalent logic. Suffice it to say that Meinong's considerations regarding the ontological foundations of probability theory, particularly as regards the status of logically simple propositions, are highly original as well as complex, and not to be dismissed lightly. Some of his ideas bore fruit indirectly through the work of a real logician, to whom we now turn.

## 7.4 Łukasiewicz on Future Contingents

The problem of future contingents is an old one, going back to Aristotle, and much discussed in medieval and modern times in relation to the theory of God's omniscience and human freedom. It had often been held that supposedly contingent events, such as (to use Aristotle's famous example) whether there will be a sea battle tomorrow, are incompatible with there being a fact or truth of the matter in advance. If, the day before the battle may take place, it is *true* that there will be a battle, then it seems there cannot *not* be a battle: a battle is inevitable, and therefore not contingent. Similar considerations apply to human freedom, the ability to do or refrain from doing something. Suppose today I am undecided how to vote in tomorrow's election, but then make my mind up and vote for candidate X, then since anyone who predicted that I was going to vote for candidate X would have been correct in their prediction, it seems as though I had no choice. Determination of truth value before the event seems incompatible with freedom and contingency.

All this is very familiar. It was discussed in Lwów, where Łukasiewicz was an associate professor, before 1914, when Lwów was still part of Austria-Hungary. Tadeusz Kotarbiński, then just completing his doctorate, wrote a paper in which he suggested future-tense propositions about such contingencies were neither true nor false,[12] while his arguments were countered by his friend and contemporary Stanisław Leśniewski.[13] So the topic was in the air in Lwów's philosophical circles. Łukasiewicz, committed to the position that humans are free, and accepting the argument that if a future-tense proposition is definitely true or definitely false, then an action or its omission is unavoidable, took Kotarbiński's solution a step further and posited a third truth value for propositions about future contingencies, which he called 'possible'. In 1917 he worked out a propositional logic with three truth values, which he published in 1920. From then onwards, many-valued logic developed rapidly and continuously at the hands of Łukasiewicz and his students.

[12] Tadeusz Kotarbiński, 'Zagadnienie istnienia przyszłości', *Przegląd Filozoficzny* 16 (1913): 74–92.

[13] S. Leśniewski, 'Czy prawda jest tylko wieczna czy też wieczna i odwieczna?', *Nowe Tory* 18 (1913): 493–528. Translation: 'Is Truth Only True Eternally or is it also True Without a Beginning?', in Leśniewski, *Collected Works* (Dordrecht: Kluwer, 1992), 86–114.

It is remarkable that Łukasiewicz developed many-valued logic not, as might be expected from its subsequent history, to deal with vagueness, or with nonsense, but to provide a vehicle for modality, the modality of the possible, lying between truth and falsity. His first two short papers on the subject, corresponding to talks delivered just two weeks apart in Lwów, are tellingly entitled, first, 'On the Concept of Possibility'[14] and second 'On Three-Valued Logic'.[15] Modern modal logics are usually bivalent, but Łukasiewicz has an argument to the effect that it is inconsistent to hang onto bivalence and also believe in possibility, as follows.

We accept [3], [4] and [5] from the first section. Assuming determinism, Łukasiewicz says we should also accept the converses of [3] and [4]:

[13] If $p$, then necessarily $p$

[14] If possibly $p$, then $p$

and if we finally accept that a proposition is possible if and only if its negation is possible:

[15] Possibly $p$ if and only if possibly not $p$

then it follows that, for all $p$:

[16] $p$ if and only if not $p$

which is always false in a two-valued logic. But if we accept a third truth value, $\frac{1}{2}$ or 'possible', then if a proposition $p$ has value $\frac{1}{2}$, so does its negation, and therefore since they have the same truth value, they are equivalent, so [16] is verified. Clearly the propositions [13]–[14] are debatable, but we simply note that and pass on.

Modal operators can now be defined as follows:

[17] For all $p$, possibly $p$ iff (if not $p$ then $p$)

[18] For all $p$, necessarily $p$ iff not possibly not $p$.

Necessity, so defined, is a strong assertion functor: $p$ is necessary if and only $p$ is neither false nor Possible.

In 1922[16] Łukasiewicz invented the first fuzzy logic, a system with infinitely many truth values between truth and falsity, represented by numbers in the interval [0,1], where 0 represents falsity, 1 represents truth, and the numbers between represent intermediate degrees of truth. Although he also developed $n$-valued logics for finite

[14] Łukasiewicz, 'O pojęciu możliwości', *Ruch Filozoficzny* 5 (1920): 169–70. Translation: 'On the Concept of Possibility', in S. McCall, ed., *Polish Logic, 1920–1939* (Oxford: Clarendon Press, 1967), 15–16.

[15] Łukasiewicz, 'O logice trójwartościowej', *Ruch Filozoficzny* 5 (1920): 170–1. Translation: 'On Three-Valued Logic', in *Polish Logic,* 16–18, and in Łukasiewicz, *Selected Works,* edited by L. Borkowski (Amsterdam, 1970), 87–8.

[16] Łukasiewicz, 'Interpretacja liczbowa teorii zdań', *Ruch Filozoficzny* 7 (1922/23): 92–3. Translation: 'A Numerical Interpretation of the Theory of Propositions', in Łukasiewicz, *Selected Works,* edited by L. Borkowski (Amersterdam: North-Holland, 1970), 129–30.

values of *n* greater than 3, in 1930 he declared that only two systems were of philosophical interest. His rationale may sound familiar:

> it was clear to me from the outset that among all the many-valued systems only two can claim any philosophical significance: the three-valued one and the infinite-valued ones. For if values other than '0' and '1' are interpreted as 'the possible', only two cases can reasonably be distinguished: either one assumes that there are no variations in degrees of the possible and consequently arrives at the three-valued system; or one assumes the opposite, in which case it would be most natural to suppose, as in the theory of probabilities, that there are infinitely many degrees of possibility, which leads to the infinite-valued propositional calculus. I believe that the latter system is preferable to all others. Unfortunately this system has not yet been investigated sufficiently; in particular the relations of the infinite-valued system to the calculus of probabilities awaits further inquiry.[17]

The uncertainty about the relationship between infinite-valued logic and probability did not go away. One thing is clear however: they cannot be the same. The reason is that the probability of a contradiction (*p* and not *p*) is always zero, whereas the truth value of *p* and *q* in Łukasiewicz's logic is the lesser of the truth values of *p* and *q*, so if *p* is neither true nor false (neither 1 nor 0), the truth value of (*p* and not *p*) is not zero but the lesser of the truth values of *p* and of not *p*; in fact if the truth value of *p* is $\frac{1}{2}$, so is that of not *p* and therefore so is that of (*p* and not *p*). Łukasiewicz's many-valued logics are extensional or truth-value-functional, that is, the truth value of a compound proposition is a function of (solely) the truth values of its components, whereas probability theory is not truth-value-functional. Hence this infinite-valued logic is unsuitable as a formal theory of probability.

Meinong's increasable possibility is not subject to this objection, because while Meinong does not stipulate anything like value-tables for his possibilities, he clearly and explicitly endorses standard principles of probability theory for compound propositions, which preclude value-functionality. Łukasiewicz's attachment to extensionality (value-functionality) is a consequence of his rather extreme view that logic is the science not of inference but of truth values:[18] it is the reason why he needs to resort to a many-valued logic in the first place to represent modality. More standard modal logics accept that propositions can have the same truth value but different modal status: a necessary truth like '2 + 2 = 4' and a contingent truth like 'Napoleon died on St Helena' for example.

[17] Łukasiewicz, 'Philosophische Bemerkungen zu mehrwertigen Systemen des Aussagenkalküls', *Comptes rendus de la Société des Sciences et des Lettres de Varsovie,* cl. iii, 23 (1930): 51–77. Translation: 'Philosophical Remarks on Many-Valued Systems of Propositional Logic', in McCall, ed., *Polish Logic,* pp. 40–65, and in Łukasiewicz, *Selected Works,* pp. 153–78. I refer here to the latter, p. 173.

[18] Łukasiewicz, 'Logika dwuwartościowa', *Przegląd Filozoficzny* 23 (1921): 189–205. Translation: 'Two-valued Logic', in Łukasiewicz, *Selected Works,* pp. 89–109. Cf. Łukasiewicz, 'Philosophical Remarks on Many-Valued Systems of Propositional Logic', in McCall, *Polish Logic,* p. 90: 'Logic is the science of objects of a specific kind, namely a science of *logical values.*'

In general it is widely held to be a drawback of Łukasiewicz's three-valued and infinite-valued logics that logical principles hitherto thought to be inviolate, namely the law of contradiction and the law of excluded middle, are not theorems, because they are on occasion not true. Łukasiewicz eventually found a way around this, to which we turn.

## 7.5 Łukasiewicz's Second Thoughts on Modality

Łukasiewicz worked for many years on a monograph on Aristotle's logic. A completed Polish manuscript was destroyed in the Second World War; when Łukasiewicz returned to the project in Ireland after the war, he rewrote his ideas in English, as *Aristotle's Syllogistic from the Standpoint of Modern Formal Logic*. The first edition of 1951 dealt only with the categorical syllogistic. It was not until the second edition, completed shortly before his death and published posthumously in 1957, that Łukasiewicz included additional chapters on the modal syllogistic. The reason for the delay was that Łukasiewicz was no longer satisfied with his three-valued modal logic, and devised a new four-valued logic, which he called simply Ł.[19]

The new logic was not the four-valued generalization of the three-valued one, but a completely new system. In it, there is not one but two values intermediate between truth and falsity, but they are incommensurable, that is, neither is stronger or weaker than the other. The new system in effect verifies principles [3]–[7], but rather than introduce quantification over truth values, Łukasiewicz prefers to work with a dualism of asserted and rejected formulas. Asserted formulas are true for all instantiations, while rejected formulas are not true for some instantiations. The quantificational equivalents of the formulas Łukasiewicz wants to reject, apart from [4] and [7], are:

[19] For all *p*, possibly *p*

[20] For all *p*, not necessarily *p*.

Neither is true because their instantiations are not true when *p* is not true.

The details of Łukasiewicz's system Ł and its peculiarities need not detain us here. Suffice it to say that while it conforms to his overall requirements, it has a number of very odd features which make it hard to accept as a modal logic.[20] These derive from two of Łukasiewicz's convictions. One is the extensionality of all logic, even modal logic, as mentioned before. The other is new, and is his dislike of necessities, as theorems or indeed even as truths. In Ł no proposition of the form 'Necessarily *p*' is true, let alone a theorem. In this it contrasts with the earlier three-valued logic, which had:

[21] Necessarily, if *p* then *p*

[19] Łukasiewicz, 'A System of Modal Logic', *The Journal of Computing Systems* 1 (1953): 111–49.

[20] G. Hughes and M. Cresswell, *An Introduction to Modal Logic* (London: Methuen, 1968), pp. 307–10; J. Font and P. Hájek, 'On Lukasiewicz's Four-Valued Modal Logic', *Studia Logica* 70 (2002): 157–82.

as a theorem. Łukasiewicz's reasoning turns on the conviction that there is no supertruth which is somehow greater than truth, that apodeictic truths are no more noble or dignified than ordinary truths. This is so similar to Meinong's position that it is hard not to surmise that they had talked about it and agreed in their views. Without this additional motivation, the original three-valued system can cope with Łukasiewicz's desiderata.

Unlike his earlier systems however, in this new four-valued system Ł, all the theorems of classical two-valued logic, such as the law of non-contradiction, remain theorems. This is because, technically speaking, the non-modal part of the system is simply the Cartesian product of classical logic with itself, and that construction automatically preserves theoremhood and valid consequence as regards the classical functors like negation, conjunction, and so on. The modal functors arise in the new four-valued context and are not available in the simpler two-valued one.

Łukasiewicz reiterates his view that apart from the new four-valued system, the only other modal system of philosophical significance is the infinite-valued system.[21] But this is an incomplete thinking-through of the revision brought about by changing the 'unincreasable' possibility (contingency) from one to two values: corresponding revisions would see either two lines of intermediate values, or an infinity of half-way intermediates and a two-dimensional infinite square of graded intermediates, with a logic at whose motivations and principles one could only guess.[22]

## 7.6 Why Łukasiewicz Could (and Should) Have Known Better

When Łukasiewicz visited Meinong in Graz, they were both working on probability. We have already discussed the output of Meinong's work, but that of Łukasiewicz emerged somewhat earlier: *Die logischen Grundlagen der Wahrscheinlichkeitsrechnung* was published in Lwów in 1913, and as the title indicates, it was written in German.[23] The short monograph already exploits the idea of a proposition which is neither true nor false, but with a difference. Propositions are allowed to be indefinite, that is, to contain variables such as the proposition '*x* is a spade', where '*x*' ranges over all the cards in a normal pack. Where, as in this case, the numbers are defined, Łukasiewicz takes the ratio *w*(*x* is a spade), which is defined as:

$$\frac{\text{the number of values of `}x\text{' making the proposition `}x\text{ is a spade' true}}{\text{the total number of values of `}x\text{'}}$$

[21] Łukasiewicz, *Aristotle's Syllogistic from the Standpoint of Modern Formal Logic* 2nd edition (Oxford: Clarendon Press, 1957), p. 180.

[22] Łukasiewicz, *Aristotle's Syllogistic from the Standpoint of Modern Formal Logic*, pp. 179–80 looks at an 8-valued system, but draws no clear conclusion apart from noting that there are in this system stronger and weaker notions of possibility.

[23] Łukasiewicz, *Die logischen Grundlagen der Wahrscheinlichkeitsrechnung* (Kraków: Spółka Wydawnicza Polska, 1913). Translation: *Logical Foundations of Probability Theory*, in Łukasiewicz, *Selected Works*, pp. 16–63.

to be something he calls the *truth value* of the indefinite proposition. Since the ratio for an indefinite proposition $p$ is a number in the range $0 \leq w(p) \leq 1$, this means he has already taken a terminological decision to use 'truth value' for numbers between 0 and 1.[24] The difference between this and the later polyvalent account is that here in 1913, definite propositions, such as 'the next card drawn is (will be) a spade', which contain no variables, have the truth value 1 or 0, depending on whether or not a spade is in fact drawn. That is unlike the later account, in which even some definite propositions, such as this one and others about future contingents, are neither true nor false.

Łukasiewicz's 1913 theory of truth values is applied to probability theory, and in this case there is no counterintuitive result regarding contradictions or other probabilistically non-independent pairs of propositions. While the truth value of '$x$ is a spade' is $\frac{1}{4}$, and that of '$x$ is not a spade' is $\frac{3}{4}$, the truth value of '$x$ is a spade and $x$ is not a spade' is not $\frac{1}{4}$, as it would be under the 1922 fuzzy theory, but 0, as is correct for probability theory. In other words, the theory is not truth-value-functional, unlike all of Łukasiewicz's later polyvalent theories.

There is a clear if slightly different precedent for Łukasiewicz's use of indefinite propositions, those containing a variable. Furthermore, Łukasiewicz was well aware of it, since it was brought to his attention by his former teacher Twardowski, and is acknowledged in section 24 of the monograph. In section 147 of his *Wissenschaftslehre* (1837),[25] Bernard Bolzano defined a concept which he called the *Gültigkeit* (validity) of a proposition with respect to determinate ideas contained in it. For Bolzano propositions are abstract or ideal entities, not sentences and not mental acts. Take a proposition such as:

[A] The card drawn by Jan at 10:55 a.m. on Sunday, 28 June 1914, is a spade.

Assuming that Jan in fact draws the Seven of Clubs, that means the proposition is false. Replacing the descriptive subject term by the standard name of the card drawn, which is a coextensive term, we get the obviously false

[B] The Seven of Clubs is a spade.

Now consider the propositions obtainable from this one by replacing the term 'the Seven of Clubs' by any of the other canonical terms for a card in the pack, e.g. 'the Two of Diamonds', 'the Ace of Spades'. Of these, thirteen are true, and 39 false, one of which is [B]. The validity of the proposition [A] is:

24 Already in 1911, Łukasiewicz stated that by 'logical values' he meant truth, falsity, and probability: Łukasiewicz, 'O wartościach logicznych', *Ruch Filozoficzny* 1/3 (1911): 52.

25 B. Bolzano, *Wissenschaftslehre. Versuch einer ausführlichen und grösstentheils neuen Darstellung der Logik mit steter Rücksicht auf deren bisherige Bearbeiter*, 4 vols, Sulzbach: J. E. v. Seidel; 2nd improved edition (Leipzig: Meiner, 1929, 1929, 1930, and 1931; reprints: Aalen: Scientia, 1970 and 1981). Critical edition: Bolzano Gesamtausgabe Series I, Vols. 11–14. English translation: *Theory of Science*, tr. Paul Rusnock and Rolf George, 4 vols (Oxford: Oxford University Press, 2014).

$$\frac{\text{the number of variants of } \textit{the Seven of Clubs} \text{ in [B] making the outcome true}}{\text{the total number of variants of } \textit{the Seven of Clubs}\text{, modulo coextensionality}}$$

and is thus 13/52 or 1/4. Notice that this figure is obtained irrespective of the fact that the actual card drawn is not a spade. If the validity of a proposition is 1, then any variant of the varied term, including the one which happens to be correct, will make its proposition true, and if the validity is 0, all variants will make the proposition false, but in neither case can we infer the converse implication. Bolzano has thus (like Meinong after him) found a way to let definite propositions be true or false while giving them a probability (here: validity) which need be neither 1 nor 0. He also went on (in section 161) to define a notion of relative validity which enabled him to deal with conditional probability. Łukasiewicz was probably made aware of Bolzano's anticipation after completing his own work, but it did not inhibit him from criticizing Bolzano for ascribing probability to definite propositions. The criticism is not well taken: Bolzano's approach may be vulnerable because of its platonism of propositions, but that is a different matter. In other respects it is exemplary.

## 7.7 Conclusion

Standard modern alethic modality concerns counterfactual scenarios, typically conceptualized in terms of alternative possible worlds. The modal theories of Meinong and Łukasiewicz, and before them, Bolzano, are couched in terms which remain closer to the real possibilities intrinsic to the actual world, which is to be expected as they were all intent on providing a logical and ontological basis for probability theory. It is attractive that they all try to merge the account of probability with the standard logic of true and false. Each approach has its drawbacks: Meinong requires non-existent objects, Łukasiewicz requires non-standard truth values, and Bolzano requires platonistic propositions and ideas. Ironically, the technically most accomplished logician, Łukasiewicz, whose invention of many-valued logic was a minor logical revolution, was the philosophically least successful, since his account of modality is hampered by his doctrinaire logical extensionalism. The instrument he fashioned has seen employment particularly in connection with vagueness, an application he did not foresee and would no doubt have deplored. It would be good to think that an account could be given which integrates the this-worldly sense of real possibility with standard logic and probability, but without the ontological overhead costs. Whether that is possible remains to be seen.

## Bibliography

Bolzano, B., *Wissenschaftslehre. Versuch einer ausführlichen und grösstentheils neuen Darstellung der Logik mit steter Rücksicht auf deren bisherige Bearbeiter*, 4 vols, Sulzbach: J. E. v. Seidel; 2nd improved edition (Leipzig: Meiner, 1929, 1929, 1930, and 1931; reprints: Aalen: Scientia, 1970 and 1981). Critical edition: *Bolzano Gesamtausgabe* Series I, Vols. 11–14. English translation:

*Theory of Science*, tr. Paul Rusnock and Rolf George, 4 vols (Oxford: Oxford University Press, 2014).

Findlay, J. N., *Meinong's Theory of Objects and Values* (Oxford: Clarendon Press, 1963).

Font, J. P. and Hájek, P., 'On Łukasiewicz's Four-Valued Modal Logic', *Studia Logica* 70 (2002): 157–82.

Hughes, G. E. and Cresswell, M. J., *An Introduction to Modal Logic* (London: Methuen, 1968).

Kotarbiński, T., 'Zagadnienie istnienia przyszłości', *Przegląd Filozoficzny* 16 (1913): 74–92.

Leśniewski, S., 'Czy prawda jest tylko wieczna czy też wieczna i odwieczna?', *Nowe Tory* 18 (1913): 493–528.

Leśniewski, S., *Collected Works* (Dordrecht: Kluwer, 1992).

Łukasiewicz, J., 'O wartościach logicznych', *Ruch Filozoficzny* 1/3 (1911): 52.

Łukasiewicz, J., *Die logischen Grundlagen der Wahrscheinlichkeitsrechnung* (Kraków: Spółka Wydawnicza Polska, 1913).

Łukasiewicz, J., 'O logice trójwartościowej', *Ruch Filozoficzny* 5 (1920): 170–1.

Łukasiewicz, J., 'O pojęciu możliwości', *Ruch Filozoficzny* 5 (1920): 169–70.

Łukasiewicz, J., 'Logika dwuwartościowa', *Przegląd Filozoficzny* 23 (1921): 189–205.

Łukasiewicz, J., 'Interpretacja liczbowa teorii zdań', *Ruch Filozoficzny* 7 (1922–23): 92–3.

Łukasiewicz, J., 'Philosophische Bemerkungen zu mehrwertigen Systemen des Aussagen kalküls', *Comptes rendus de la Société des Sciences et des Lettres de Varsovie*, cl. iii, 23 (1930), 51–77.

Łukasiewicz, J., 'A System of Modal Logic', *The Journal of Computing Systems* 1 (1953): 111–49.

Łukasiewicz, J., *Aristotle's Syllogistic from the Standpoint of Modern Formal Logic* 2nd edition (Oxford: Clarendon Press, 1957).

Łukasiewicz, J., *Selected Works*, edited by L. Borkowski (Amsterdam: North-Holland, 1970).

McCall, S., ed., *Polish Logic, 1920–1939* (Oxford: Clarendon Press, 1967).

Meinong, A., *Über Möglichkeit und Wahrscheinlichkeit. Beiträge zur Gegenstandstheorie und Erkenntnistheorie* (Leipzig: Barth, 1915; Reprinted Graz: Akademische Druck- u. Verlagsanstalt, 1972).

Russell, B., 'Meinong's Theory of Complexes and Assumptions', *Mind* 13 (1904): 204–19, 336–54, 509–24.

Russell, B., *The Collected Papers of Bertrand Russell* vol. 4., edited by A. Urquhart (London: Routledge, 1994).

Simons, P., 'Über das, was es nicht gibt: Die Meinong–Russell Kontroverse', *Zeitschrift für Semiotik* 10 (1988): 399–426.

Simons, P., 'Łukasiewicz, Meinong, and Many-Valued Logic', in *The Vienna Circle and the Lvov-Warsaw School*, edited by K. Szaniawski (Dordrecht: Kluwer, 1989), 249–91.

Simons, P., *Philosophy and Logic in Central Europe from Bolzano to Tarski* (Dordrecht: Kluwer, 1992).

# 8

# Heidegger on 'Possibility'

*Mark Sinclair*

Martin Heidegger's project in *Being and Time* (*Sein und Zeit,* 1927)[1] involves a critique of a form of 'actualism' in philosophy together with the promotion of a certain idea of possibility. This first emerges in the remarks in §7 of the text concerning the idea of phenomenology as a school or method in twentieth-century philosophy: the 'essential character' of phenomenology, Heidegger writes, 'does not consist in its actuality as a philosophical "movement". Higher than actuality stands *possibility* (Möglichkeit)' [SZ 38]. Heidegger thus seems to advance the doubtless difficult thought that the possibility of something, here the phenomenological school, is more proper to what or how it is than its actuality. The thought is advanced more deliberately later in the text: 'possibility', Heidegger writes in §31, 'is the most primordial and the ultimate (*ursprünglichste und letze*) positive ontological determination' [SZ 143] of being—of, first of all, the being (the *Dasein*, in Heidegger's German) that each one of us is. Far, then, from having merely a methodological significance within a reflection on the idea of phenomenology, a notion of possibility as somehow constitutive of the essence of being is, for Heidegger, the base and the summit, the alpha and omega of ontology. Possibility, on this account, is not distinct from being, and it does not constitute a realm of possibilia that is not quite, not yet or not fully in being; it rather belongs to the essence of being itself, and it can do so because being, for Heidegger, is not to be equated with traditional ideas of 'actuality'. Whatever else the text of 1927 has to say about the meaning of being—with its most fundamental task consisting of showing how time is the 'horizon of any understanding of being whatsoever' [SZ 1]—at the very heart of *Sein und Zeit* stands a reflection on *Sein und Möglichkeit,* on being and possibility.

In the lecture course of the winter semester 1925–26, *Logic: The Question Concerning Truth,* Heidegger had signalled the importance of an idea of possibility for his

[1] I refer to the fifteenth edition of *Sein und Zeit* (Tübingen: Max Niemeyer, 1984) in square brackets in the body of the text as SZ. Both available English translations of the text contain the pagination of the German edition as marginalia, and thus I do not refer to them. I am indebted to Matt Barnard and Joseph Carter for responses to drafts of this essay.

philosophical project. He even suggests that his task consists in clarifying the nature of possibility as such:

[t]he concept of possibility is quite obscure [*ganz ungeklärt*] in scientific philosophy hitherto; and the extent to which it is clarified is normally limited to possibility in the sense of modality, of modality which is seen in the context of statements and their possible certainty. In this way, the idea of possibility is bound up with actuality and necessity as determinations of being, and indeed of the being of nature in the widest sense. The meaning of possibility and the type of structures of possibility belonging to *Dasein* as such have remained wholly concealed from us up to the present day.[2]

These remarks contain two clues for understanding Heidegger's project in the 1920s. Heidegger signals, first, that he is not offering a conception of possibility within the traditional parameters of a doctrine of modality insofar as he is not directly concerned with the modality of statements and their truth conditions. Even a distinction of *de dicto* and *de re* modality, of modality 'in language' and 'in things', is of little avail in order to understand Heidegger's approach insofar as both are asserted and understood *through* language. Heidegger, as will become clear, seeks to bring to light an understanding of possibility in things that is prior and perhaps irreducible to any propositional and theoretical stance we take towards the world. Some care is required with this idea of possibility 'in things', however, since Heidegger also signals that he is concerned less with the modal status of the things that we are not—'nature in the widest sense'—than with the being that we are. It is by focusing—though not exclusively, as will become apparent—on human being, on the 'structures of possibility belonging to *Dasein*', that Heidegger aims to grasp a sense of possibility as constitutive of the very essence of being, and thus to rethink the very idea of possibility.

This emphasis on an idea of possibility does not, however, end with the *Daseinsanalytik* of *Being and Time* and Heidegger's project of 'fundamental ontology' in the 1920s. It is equally essential to his later work. According to *Contributions to Philosophy*, a text written between 1936 and 1938 that is often held to constitute Heidegger's second major work, it is precisely by means of a notion of possibility that 'another beginning' [*ein anderer Anfang*] in philosophy can be instituted: 'the *possible* [*das Mögliche*] essentially occurs in being [*Seyn*] alone and as its deepest fissure, so that in the thinking of the other beginning being must first be thought in the form of the possible'.[3] This other beginning in philosophy, which takes its lead

[2] Martin Heidegger, *Gesamtausgae vol. 21: Logik: Die Frage nach der Wahrheit*, edited by W. Biemel (Frankfirt am Main: Klostermann, 1995), p. 228; *Logic: The Question of Truth*, trans. T. Sheehan (Bloomington: Indiana, 2010), p. 191. After having initially provided a full bibliographical reference to a volume of Heidegger's *Gesamtausgabe*, and to its English translation, I refer to it with the abbreviation GA followed by the volume number, page number, and, after a forward slash, the page number of the translation. I have often modified the translations, as in the passage cited: translating *ganz ungeklärt* as 'wholly unclarified', instead of 'quite obscure', makes Heidegger contradict himself in the remainder of the sentence.

[3] Martin Heidegger, *Gesamtausgabe vol. 65: Beiträge zur Philosophie* (*Vom Ereignis*) (Frankfurt am Main: Vittorio Klostermann, 1994), p. 475; *Contributions to Philosophy (Of the Event)*, trans. R. Rojecwicz and D. Vallega-Neu (Bloomington: Indiana University Press, 2012), p. 374.

from an idea of the possible, is necessary, Heidegger contends, precisely because of the predominance of ideas of actuality in the history of metaphysics. 'Metaphysics', as he writes, using the term in a pejorative sense, 'makes the "actual" as what is [*als das Seiende*] its starting point and the goal of any determination of being', whereas a more fundamental and original thinking of being will apprehend being as the possible.[4]

A notion of possibility, then, is central in Heidegger's thinking, both within his 'fundamental ontology' of the 1920s and after the *Kehre* or turn that marks his philosophical development in the 1930s. Yet how exactly are we to understand this or these conceptions of possibility, and how exactly are we to understand them in relation to traditional doctrines of modality?[5] A principal aim of the present chapter is to show how Heidegger's account of *Dasein's Möglichsein* or being-possible relies on an interpretation of its being as a form of movement. Heidegger is able to consider *Dasein*'s being as being-possible because he considers *Dasein*—like the phenomenological school of thought, as we saw above—as a movement. The idea of *Dasein's* movement or *movedness* [*Bewegtheit*] is relatively underdeveloped in the text of *Being and Time*, but it emerges, as I will show, in and from the interpretations of Aristotle's conception of modality and movement that Heidegger had advanced earlier in the decade. Although the declarations in *Being and Time* concerning the superiority of possibility in 1927 contradict Aristotle's statements concerning the superiority of actuality or *energeia*,[6] previously, before his explicit formulation of the project of 'fundamental ontology' according to the conjoined questions of *being* and *time*, Heidegger had attempted to retrieve a conception of being as *Möglichsein* from the Stagirite. If one can justifiably claim that there are 'in Western thought, three broad conceptions of possibility',[7] Heidegger is concerned neither with a doctrine of possibilia as distinct from the actual world, nor with a critical theory of modality in a Kantian sense, but rather with—though he aims to radicalize

[4] Heidegger GA 65 475/374.

[5] Attempts to address Heidegger's conception of possibility directly and to relate it to traditional doctrines of modality have been, perhaps surprisingly, rare in Heidegger studies. Wolfgang Müller-Lauter's *Möglichkeit und Wirklichkeit bei Martin Heidegger* (De Gruyter: Berlin, 1960) is the longest and most direct study, but it does not stand as an exception to the rule that anything published on Heidegger prior to the 1980s, given the publication of his *Gesamtausgabe*, is of merely historical interest. Without referring to Müller-Lauter's work, in his 'Heidegger, The Possible and God' (first published in *Heidegger et la question du dieu*, edited by R. Kearney and J. O'Leary, Paris: Grasset, 1981; republished in Heidegger, *Critical Assessments* vol. 4, edited by C. McCann, London: Routledge, 1992, 299–324), Richard Kearney noted that the question of the possible had 'hitherto been much neglected by Heidegger's commentators' (p. 299), and addresses it directly both in the essay and in his *La poétique du possible: phénoménologie herméneutique de la figuration* (Paris: Beauchesne, 1984). Of more recent scholarship, William McNeill's 'Rethinking the Possible: On the Radicalisation of Possibility in Heidegger's *Being and Time*' (in *The Condition of Possibility, theory@buffalo* 13 (2009), 105–25), offers insightful remarks on Heidegger's thinking, but has a narrower focus than this essay and is less concerned with Heidegger's situation in the history of philosophy.

[6] See Aristotle, *Metaphysics* IX, 1049b13.

[7] J. N. Mohanty, 'Husserl on "Possibility"', *Husserl Studies* 1 (1984): 13–29, p. 21.

it—possibility in an Aristotelian sense of potentiality as an ontological determination of things that is the condition of, and contrasts with, their actuality.

The second section of the essay addresses Heidegger's account of *Dasein* as a being-possible in terms of its movedness, while the third shows how this movedness is ultimately to be thought as the movement of time—time in the particular sense in which Heidegger accounts for it, namely as *temporality*. The fourth section then shows how a proper grasp of *Dasein's* movedness as a being-possible allows us to understand *Being and Time's* controversial analysis of death as the 'possibility of the impossibility' of existence. Yet the scope of this paper is limited neither to Heidegger's account of *Dasein* as a being-possible nor to his philosophical project in the 1920s. For after having examined, in the first section of the paper, the modal sense of the account of tool-being in *Being and Time*, the fifth section is concerned to show how Heidegger's reflection on the 'modality' of art production in the 1930s introduces a shift in his interpretation of possibility and his interpretation of Aristotle's modal thinking in a way that is pivotal for his *Kehre*.

## 8.1 Handiness and Being-Possible

In *Being and Time* possibility is both a 'category (*Kategorie*)' and an 'existential (*Existentiale*)', which is to say that it characterizes the being of things as well as the being—the *Existenz*—of *Dasein*, the being that we are. Although Heidegger does not dwell on this dichotomy, it is important to recognize it for, as I will show, it is a changed conception of possibility in relation to the things that we are not that is at the heart of his *Kehre* in the 1930s. The brief remarks on possibility as a category in §31 of *Being and Time*, however, presuppose the famous analysis of tool-being or handiness earlier in the text. §§15–18 advance the claim that prior to being the isolated objects of a disinterested theoretical gaze, things show themselves as pointing beyond themselves within the horizon of my practical concerns. Things are apprehended as 'useable for', 'good for' a particular purpose, and in the given situation each thing is seen in relation to others: the hammer, for example, points beyond itself to the nails and to the boards within the horizon of the task at hand. Things in their individuality withdraw themselves from my attention to the degree that they are used, to the degree that I am absorbed by my practical project, but, for Heidegger, this observation has ontological and not merely psychological significance. Things encountered within the horizon of my practical concerns are *zuhanden*—their being, in other words, is not objectivity, or an indeterminate notion of 'reality in general', but rather *Zuhandenheit*, being-ready-to-hand. Being-ready-to-hand is not simply a property of something, of something *vorhanden*, as Heidegger puts it—which is to say, explicitly present as an object.[8] It is still less the product of a merely

[8] The sense of the reference to the *hand* in the term *Vorhandenheit*—which for Heidegger serves to translate *existentia* [SZ 42]—is clarified only by Heidegger's 'destruction of the history of ontology', an essential element of his project of fundamental ontology. One aspect of this 'destruction' concerns the way

subjective judgement or valuation. Being-ready-to-hand is rather a pre-thematic purposiveness that is intrinsic to things, and which determines *how* they exist, their way of being.

Certainly, this intrinsic purposiveness does not occur without the practical project of the agent; the thing cannot be purposeful without someone with a purpose. According to §31, the purposiveness of the thing is a function of *Dasein*'s pre-predicative, pre-conceptual understanding, which constitutes the horizon in which things appear. This practical horizon is one aspect of what Heidegger understands as 'world', which is not a thing or a collection of things, and rather belongs to *Dasein*'s being as *in-der-Welt-sein,* being-in-the-world.[9] The understanding of this practical horizon of world is no mere passive reception of the given, but a projection or *Entwurf* that structures the agent's dealings with particular things. Yet projection here is not to be understood in the sense of a secondary, and ultimately fictive interpretation of intrinsically purposeless things *vorhanden.* Instead, the pre-thematic purposiveness at once understood and projected by the agent determines, on Heidegger's account, the very being of things ready-to-hand. In our every comportment towards beings, Heidegger contends, there is an understanding of these beings in their being, and in engaging with what is ready-to-hand, there is and must be an understanding of their *being*-ready-to-hand.

In §18 Heidegger had discussed *Dasein's* encountering of things ready-to-hand within the horizon of a practical project as a *Freigabe*, as a making-free or freeing-up of the thing for what it is good for; things, insofar as I engage with them, are freed up to be what or, better, how they are, namely ready-to-hand. However, in §31, which contains some of *Being and Time's* most programmatic remarks on possibility, he accounts for this freeing-up, briefly but no less emphatically, in modal terms: 'when that which is in the world is itself freed, this entity is freed for its own possibilities. The ready-to-hand is discovered as such in its service*ability*, us*ability*, detriment*ality* [Dien*lichkeit*, Verwend*barkeit*, Abträg*lichkeit*]' [SZ 145]. Within the horizon of my practical projects things are encountered as useful, useable, available, or, on the contrary, as unavailable or as detrimental, and this practical possibility, this form of practical modality, does not just reside in the agent's thoughts—in my, as Kant would have it, merely subjective teleological judgements. There is an awareness of modality prior to explicit conceptual thought, and this practical modality belongs as much to the things as it does to the person using them. To be sure, the awareness of this practical modality is not gained by 'the theoretico-thematical consideration of the possible as possible, and by having regard for its possibility as such', but rather by a concern for what I can do and make actual with the instrument in hand, by 'looking

in which being in the philosophical tradition means being-produced, being-(hand)made. On this point, see chapter 1 of my *Heidegger, Aristotle and the Work of Art* (Basingstoke: Palgrave, 2006).

[9] See §15 of *Being and Time.*

circumspectively away from the possible and looking at that for which it is possible [*das Wofür-möglich*]' [SZ 261]. In no way do I need to be reflectively aware of the ready-to-hand for it to be ready-to-hand; on the contrary, the ready-to-hand as such precedes and to a certain degree escapes explicit, conceptual awareness.

The basic point, for us, is this: there is a practical consciousness, an 'I can', that underlies and precedes the reflective self-consciousness of the 'I think', but the 'I can' is given and coeval with an 'it can', through the pre-reflective possibilities afforded to me by the thing or things in question. Now, other thinkers in the phenomenological tradition, including Edmund Husserl, may well offer versions of this insight, and we can gain a better grasp of the specificity of Heidegger's approach in comparing it to that of his teacher. Husserl famously argues that in perceptual experience the present thing is given with a kind of halo or horizon of potentialities; part of the intended sense of my perception of, say, this table here and now, and from the position I perceive it, is that when I move other presently invisible aspects of the table will present themselves. The unity of the three-dimensional object in experience is not, in other words, as the thoroughgoing empiricist will contend, a product of mere associative and secondary processes in the mind. In this sense, in describing the 'Actuality and Potentiality of Intentional Life', Husserl claims that 'every actuality involves its potentialities, which are not empty possibilities, but rather possibilities intentionally delineated';[10] other possible aspects of the table are delineated in the aspect of the table that I see, and these possibilities, since they constitute an aspect of the intended sense of any given object, are more determinately rooted in perceptual experience than any mere logical possibility. It can be argued, following J. N. Mohanty, that this horizon belonging to all objective experience is not merely a matter of intellectual cognizance, and that it is in and of itself a 'practical horizon... indicating a system of possibilities for practical intervention'; the table presents itself as something that I *can* work on or walk around, and in this way the 'pre-delineated possibilities of fulfilment are practical possibilities'.[11] The potentialities given in perceptual consciousness are, from the ground up, a function of a practical consciousness, of an 'I can'. Understood in this way, Husserl's account of the horizons in and of perception stands in the closest proximity to Heidegger's account of *Zuhandenheit*. There is nevertheless an essential difference in their approaches: for Heidegger, it is not simply the case that the *actual* thing is present within an ultimately practical horizon of pre-delineated possibilities or potentialities. The very being of the thing, Heidegger urges us to recognize, is something other than actuality or *Vorhandenheit* precisely insofar as the thing recedes from conscious awareness as an isolated object when I purposefully go about my projects. In short, Husserl sees individual things with a halo or shadow inviting a practical response, whereas Heidegger

[10] Edmund Husserl, *Cartesian Meditations* (The Hague: Martinus Nijhoff, 1960), §19, p. 45.
[11] J. N. Mohanty, 'Husserl on 'Possibility'', pp. 27–8.

sees things as intrinsically and pre-thematically interrelated, each pointing beyond itself within the teleological horizon of the given situation.

One might wonder whether in his account of tool-being Heidegger exaggerates the extent to which things are pre-objective or non-thematically apprehended in practical experience. One might also wonder why he begins with tools and the world of the workshop in a narrow sense in order to account for practical experience in general.[12] For our present purposes, however, it is sufficient to note that an account of this pre-thematic projective understanding of possibility is no mere ancillary detail in a theory of modality. On the contrary, it constitutes a fundamental awareness of modality, an awareness given prior to the explicit grasp of conceptual possibility, i.e. to concepts of what could be objectively present. This particular form of modal understanding is prior to conceptual possibility, just as, on Heidegger's account, things are first encountered as *zuhanden* before their possible apparition as objectively present, as *vorhanden*. Yet this epistemological precedence is accompanied by an ontological superiority. In the case of possibility as 'modal category of *Vorhandenheit*', Heidegger writes, 'possibility means what is *not yet* actual and what is *never* necessary. It characterises what is merely possible. Ontologically it is on a lower level than actuality and necessity' [SZ 143]. We might understand possibility in this sense as 'mere empty logical possibility' [SZ 143], or in a more real or metaphysical sense, following Kant, as characterizing that which accords with the—transcendental—conditions of experience, but in either case it characterizes a deficient mode of being. Within Heidegger's analysis of tool-being, in contrast, possibility determines the fullest and most original mode of the being or existence of things. Traditional doctrines of modality have passed over this sense of possibility; and they have passed over it precisely because of the predominance of an idea of existence understood—in different ways, certainly, at different moments of the tradition—as objective presence. For Heidegger, throughout the tradition the modal categories—actuality, necessity, possibility—are modes of objective presence, i.e. of existence or actuality, with one of the modes standing thus as the measure of the other two, or else all three are taken as modes of 'something' other, i.e. of being, the meaning of which has never been adequately brought into question.

On this basis, and even before examining Heidegger's analysis of *Dasein* as a being-possible, we gain a preliminary understanding of the stakes of his critique of actualism in philosophy. 'Actualism' here does not simply signify the doctrine according to which only actual things—the table, say, that I am writing on—exist, as opposed to possible things—the unicorn, say, I am thinking about—which, claims the actualist, have no being at all. From Heidegger's perspective, 'actualism', more fundamentally, amounts to the idea that being or existence is identical to actuality, i.e. to objective presence. Actualism in this more fundamental sense is, in

[12] On both these questions, see the second chapter of my *Heidegger, Aristotle and the Work of Art*.

the end, the prior ground of actualism in the narrower, contemporary sense, i.e. of attempts to exclude possible things from the realm of being: it is by assuming that being means objective presence that possibles are excluded from its domain as a result of being insufficiently objective or present. This is not to say, of course, that Heidegger is on the side of the 'possibilists' within contemporary debates in the metaphysics of modality. He is little concerned with the status of possibilia, with possibility in the sense of possibly objectively present, i.e. possibly actual things.[13] He rather commits his entire philosophical career to the thought that the meaning of being cannot, or at least should not, be restricted to objective presence—and instead of wondering if and to what extent possibilities are actual, Heidegger urges us to question the predominance of ideas of actuality in metaphysics.

## 8.2 *Dasein*'s Movedness as Being-Possible

Heidegger points to the peculiar modal status of being-ready-to-hand almost in passing in §31, and his more basic concern in this section is to elucidate, at least provisionally, the being of *Dasein* as a being-possible. *Dasein* understands the particular possibilities afforded to it by things encountered pre-objectively, yet it does this only against the background of an understanding of its own projects, projects that are but possibilities of its own being.[14] I can choose to do one thing or another, and this means that I can choose to become, as is sometimes said, one particular person or another—hero or traitor, stoic or coward, dissolute or disciplined. Yet Heidegger leads us away from the idea that we are simply something or someone with possibilities, to the idea the very being of *Dasein* is a being-possible. The possibilities that *Dasein* has are not to be thought of as a present-at-hand quality or attribute of something—the 'person'—also present at hand. §9 of *Being and Time* had already suggested that *Dasein*'s way of being is irreducible to any traditional notion of existence—to any conception of *actualitas* or *existentia*—and this precisely because it is characterized by possibility:

> [t]hat entity which in its Being has this very Being as an issue, comports itself towards its Being as its ownmost possibility. In each case *Dasein is* its possibility, and it 'has' this possibility, but not just as a property, as something *vorhanden*. [SZ 42]

*Dasein* is a being concerned for what or how it can possibly be; and this possibility is 'ownmost' or most proper to *Dasein* in the sense that no one else can live out this concern, i.e. live my life, for me. In its concern for how it can possibly be, *Dasein* is possibility from, as it were, the ground up.

[13] Michael Inwood notes that Heidegger is not directly concerned with any form of modality in a logical sense, and he 'has no more interest in logical necessity than in logical possibility'; M. Inwood, *A Heidegger Dictionary* (Oxford: Blackwell, 1999), p. 172.

[14] As Heidegger puts it, understanding 'projects the being of *Dasein* with respect to that for the sake of which it exists with equal primordiality as it projects *Dasein*'s being with respect to the significance that constitutes the worldliness of a particular world' [SZ 145].

§31 attempts to address directly, then, the issue of *Dasein*'s *Existenz* as possibility announced earlier in the text, but even here Heidegger announces that this issue can be 'prepared as a problem' [SZ 144] only. We should not be surprised by this: if possibility in some sense 'is' *Dasein*'s being, and if—as Heidegger recalls in an Aristotelian fashion within the opening pages of *Being and Time*—being cannot be defined [SZ 3], then possibility too will not allow itself to be captured in the form of a simple definition. Moreover, if possibility in some sense is being, then an account of the nature or essence of possibility will at least partially depend on what Heidegger has to say in Part II of the text of 1927 about being and time, about 'time as the horizon for any understanding of being' [SZ 1]. The first steps in the preparation of possibility as a problem in §31, however, involve an elucidation of what possibility in this sense *is not*. It is not—as we have already seen in discussing possibility as a category—mere logical possibility, or a modal category subordinate to actuality understood as *Vorhandenheit*. It is also to be distinguished from contingency, from the non-necessity characterizing the being that can change, that can come into and go out of existence.

As Heidegger additionally, and crucially, remarks, possibility as an *existentiale* is not to be taken as the 'free-floating object of a purported *liberty of indifference* (*libertas indifferentiae*)' [SZ 144]. *Dasein*'s possibilities are not the object of an arbitrary or indifferent choice, in the way that one chooses a main course from a menu when unmoved by any of the options. To conceive possibility in this way would be to misconceive both the nature of *Dasein* as a being that in some sense 'chooses' and the nature of the possibilities from which it 'chooses'.[15] *Dasein* does not sit in judgement on objective possibilities that are simply indifferent to it, and it does not survey these possibilities from a position external to them; on the contrary, it always and already finds itself in a world, with a history, and thus as already having taken up definite possibilities:

> In every case, *Dasein*... has already got itself into definite possibilities. As the potentiality-for-being (*Seinkönnen*) which it *is*, it has let such possibilities pass by; it is constantly waiving the possibilities of its being, or else it seizes upon them and makes mistakes. But this means that *Dasein* is being-possible (*Möglichsein*) which has been delivered over to itself—*thrown possibility* through and through. [SZ 144][16]

[15] For a recent discussion of 'choice' as irreducible to intellectual deliberation in Heidegger's text, see Béatrice Han-Pile, 'Freedom and the "Choice to Choose to Oneself"' in *The Cambridge Companion to Being and Time*, edited by M. Wrathall (Cambridge: Cambridge University Press, 2013), 291–319.

[16] Passages like this suggest that Heidegger aims in some sense to distinguish *Dasein*'s potentiality-for-being or ability-to-be (*Seinkönnen*) from its 'being-possible' (*Möglichsein*). It is far from obvious, however, that Heidegger is attempting to mark the difference between 'our life projects, on the one hand, and our projecting ourselves into those projects, on the other', as Iain Thomson claims without elucidating or substantiating his claim in any way, in 'Death and Demise in *Being and Time*', in *The Cambridge Companion to Being and Time*, edited by M. Wrathall (Cambridge: Cambridge University Press, 2013), 260–90, p. 269.

*Dasein's* understanding of possibilities is certainly a function of a projection or *Entwurf*, but this projection is itself always and already *projected* or *thrown* [*geworfen*] in that we always and already finds ourselves in a given situation, with a world and at a particular point in history. *Dasein* is bound to this world and history and dependent on it. Possibility in some sense constitutes the essence of *Dasein*'s freedom, but according to this idea of throwness, *Dasein*'s freedom is not absolute; *Dasein*, to be sure, is not autonomous in the sense of self-grounding.[17]

Yet if *Dasein* is not an ahistorical, isolated, self-grounding subject, then just as little are the possibilities it 'chooses' objects for it:

> ... the character of understanding as projection is such that the understanding does not grasp thematically that upon which it projects—that is to say, possibilities. Grasping it in such a manner would take away from what is projected its very character as a possibility, and would reduce it to the given contents which we have in mind. [SZ 145]

The possibilities offered by things can be understood conceptually and reflectively; but prior to this, *Dasein* has an understanding of a different strata of possibilities, which ultimately are possibilities of its being, and which are pre-thematic, pre-predicative and pre-conceptual. This priority constitutes, again, an ontological superiority rather than deficiency; possibility in this sense, according to *Being and Time*, is possibility in the most genuine sense. Heidegger even seems to argue that an intellectualist construal of possibility—i.e. understanding possibility as conceptual or ideal—would reduce possibility to actuality, for conceptual possibilities, though not actually present in the world, are nevertheless concepts of possibly actual things or events.

Heidegger may well be gesturing here at a combinatorial construal of possibility: conceptual possibilities are, in the end, actualities, because these concepts, supposedly like all concepts, derive from sense-experience. For the combinatorialist, unicorns, though not actual, are possible precisely and only insofar as their idea is combined from those of actual horses and horns. If this is Heidegger's intention, his approach, it is worthwhile to remark, shares common ground with the account of possibility that Henri Bergson, a thinker to whose conception of time as duration Heidegger is evidently indebted, was developing at around the same time. In his '*Le possible et le réel*',[18] Bergson argues, in endorsing a traditional identification of possibility with conceivability, that the possible does not precede the real, as, say, Leibniz had it, but rather follows from it, since our ideas of what can be possible derive only from reality. On this basis, Bergson offers a particular and radical response to the oft-invoked difficulty of accounting for novelty within a combinatorial construal of possibility: things or events, in their novelty, *are not possible before they occur*. *Macbeth*,

[17] On this point, see section III of William McNeill's 'Rethinking the Possible: On the Radicalization of Possibility in Heidegger's *Being and Time*'.

[18] It was published in his 1934 volume *La pensée et le mouvant*, recently re-edited by A. Bouaniche et al (Paris: Presses Universitaires de France, 2009).

say, was not possible before it was written precisely because it was not foreseeable, i.e. conceivable before it was written; and it became possible, i.e. conceivable, only as an actually existing work of art. Now, Heidegger might accept many of the elements of this particular interpretation of possibility as conceivability, but, in distinction to Bergson, he aims to think under the heading of 'possibility' a more fundamental sense of modality that is irreducible to conceivability.[19]

According to *Being and Time,* in any event, *Dasein* is not distinct from the pre-conceptual possibilities that it projects and understands, for these possibilities are, at bottom, possibilities of its own being. Yet is not enough to say that *Dasein* is not distinct from the possibilities that it projectively understands, for Heidegger's fundamental and more positive claim is that *Dasein is* the possibilities that it projects:

> ...projection, in throwing, throws before itself the possibility as possibility, and lets it *be* as such. As projecting, understanding is the kind of being of *Dasein* in which it *is* its possibilities as possibilities. [SZ 145]

*Dasein* already *is* its possibilities and thus these possibilities do not simply constitute an imperfect state *from which Dasein* moves in order subsequently to become the *Dasein* that it really is. To claim this is not, to be sure, to deny that Heidegger aims to account for *Dasein*'s being according to an idea of movement. *Being and Time* is certainly—even though Heidegger says little explicit about this in the text of 1927—grounded on the idea that *Dasein* has a form of movement all of its own, a 'movedness (*Bewegtheit*)' [SZ 374] that is analogous but irreducible to locomotion or to any other Aristotelian category of movement or change applicable to things. The characterization of the 'kind of being' proper to *Dasein* in the passage cited above recalls, in fact, Aristotle's 'definition' of movement in *Physics* III as the 'actuality of the possible as such (*tou dunamei ontos entelecheia hei toiouton*)'—as the actuality of the possible as possible.[20]

In order to understand Heidegger's appropriation of Aristotle in this connection, it is crucial to see that the Stagirite's definition of movement can hardly be held to account for it simply as the *transition* from potentiality to actuality. Such a transitional account of movement would be insufficient in at least two ways: first, it would be circular, since it accounts for movement as a transition, i.e. as a movement

[19] In *La poétique du possible*, p. 35, Richard Kearney thus rightly notes that, from Heidegger's perspective, we have to distance ourselves from Bergson's conception of possibility. Felix O'Murchadha's contrasting claim (*The Time of Revolution: Kairos and Chronos in Heidegger,* London: Continuum, 2013) that Bergson and Heidegger 'share the thought' that 'if we are to understand possibility on the basis of freedom, then it can no longer be thought as a realm of present options which can be chosen' (p. 24) is unhelpful in that it goes beyond anything that Bergson actually says about possibility. On the question, however, of how Bergson's philosophy does require a more positive sense of modality than the critique of possibility he presents in 1934, see my 'Bergson on Possibility and Novelty', *Archiv für Geschichte der Philosophie* 96/1 (2014): 104–25.

[20] Aristotle, *Physics* III.1, 201a10. That Heidegger rehearses Aristotle's definition of motion in the passage last cited from §31 of *Being and Time* seems to have escaped the notice of even those commentators most concerned to stress the importance of Aristotle's definition of motion for Heidegger's *Daseinsanalytik.*

between two states; and, second, it tells us nothing about movement itself and only something about the states between which the thing in movement moves.[21] We have good reasons to consider that Aristotle is thinking of neither a process nor the result of a process, but rather of a particular mode or way of being—*movement*—in which the possible really or genuinely exists as the possibility that it is. This is the way that Heidegger interprets Aristotle's definition in a lecture course of 1924 on the *Basic Concepts of Aristotelian Philosophy*:

> movement is the *entelecheia*, the presence [*Gegenwart*] of beings *as* the ability of being-there [*als des Daseinkönnen*], and indeed this presence *as long as* it is able to be there. Motion is the presence of the ability of being-there as such.[22]

Movement is 'where' possibility exists fully as possibility in the sense that the potentiality of the wood to form a statue only really becomes apparent, and only fully exists, in the process of its realization. Movement, as Aristotle says, is a certain type of *energeia*, of being-in-work or activity, an activity that is *ateles*,[23] which has not yet come to its end, and it is insofar as the movement has not finished that the possible can genuinely appear as the possible. Of course, given that the realization of a possibility is, in the case of a movement, the abolition of that possibility, Heidegger is also claiming that possibility genuinely appears only on the way to its abolition. Despite its air of paradox, Aryeh Kosman has more recently advanced a similar interpretation: Aristotle's definition of movement attempts to reveal 'the *activity* of being able to be',[24] an activity that does not yet characterize the idle potentiality of the wood to form a statue and that is no longer possessed by the statue as a finished product.

In order to understand Aristotle's definition of movement, it is crucial to see that there are levels of potentiality: the potentiality of the wood to form a statue is latent and inactive when the wood is not being worked on, but manifest as an 'active potentiality', a *tätige Möglichkeit*,[25] when the word undergoes change by means of the work of the craftsman. Movement or change in the widest sense is certainly

[21] For these arguments, see Aryeh Kosman, 'Aristotle's Definition of Motion', *Phronesis* 14/1 (1969): 40–62, p. 42.

[22] Heidegger, *Gesamtausgabe vol. 18: Grundbegriffe der aristotelischen Philosophie* (Frankfurt am Main: Vittorio Klostermann, 2002), p. 313; *The Basic Concepts of Aristotelian Philosophy*, trans. R. Metcalfe and M. Tanzer (Bloomington, IN: Indiana University Press, 2009), p. 211. In 1924, *Dasein* is not yet a term of art in Heidegger's philosophical lexicon, and here signifies existence in general.

[23] Aristotle, *Physics* V.2, 201b32.

[24] Aryeh Kosman, *The Activity of Being: An Essay on Aristotle's Ontology* (Cambridge, MA: Harvard University Press, 2013), p. 68. Kosman acknowledges a kinship with Heidegger.

[25] Heidegger writes this in his *Handschriften* to the lecture course of the summer semester 1924: GA18 378/256. In his 'Heidegger's *Sein zum Tode* as Radicalization of Aristotle's Definition of *Kinesis*', *Epoché* 18/2 (2014): 473–502, by showing that Heidegger is offering an account of levels of potentiality in Aristotle, Joseph Carter easily rebuts Francisco Gonzalez's claim ('Whose Metaphysics of Presence? Heidegger's Interpretation of *Energeia and Dunamis*', *The Southern Journal of Philosophy* 44 (2006): 533–68) that Heidegger offers a muddled interpretation of Aristotle's concept of movement in the lecture course of 1924.

movement *from* something *to* something, from one state to another, from potentiality to actuality, but it is the peculiar being of this from-to structure, of this *being on the way to completion, a being on the way where possibility fully exists as possibility*, that is the focus of Heidegger's interest in Aristotle's conception of movement.[26] It is the peculiar 'presence (*Gegenwart*) of this being-from-to (*Von-zu-Sein*)'[27] structure that is at issue, as Heidegger claims in 1924—and it is this that he will incorporate within an account of the movedness proper to *Dasein*. *Dasein* is somehow stretched out between its possibilities and their realization. Indeed, as Heidegger will put it in Part II of *Being and Time*, the movedness of *Dasein* is a function of a 'stretching', which is necessarily a '*stretched* out *self-stretching* [*erstreckten Sicherstreckens*]' [SZ 374–5] since there is no external agency which stretches *Dasein* out. It is insofar as *Dasein* is stretched out in this way that, *on the one hand*, its possibilities can fully and genuinely exist as possibilities, and that, *on the other hand*, it can exist as these possibilities. In existing, *Dasein* certainly moves from particular possibilities to their realization, but the being of *Dasein*—as a being that is always in 'movement', as a being that is not, for as long as it is alive, a 'finished product'—consists in the peculiar stretched-out being or activity of the possible that *is* its movement.

Although *Dasein* can realize particular possibilities, it can never simply be those possibilities actualized, for it is continually in movement, continually on the way to another possibility of its own being. *Dasein* can *be* the possibilities that it understands and projects, possibilities that it *is not* yet—but it can never simply be the possibilities once actualized since it is always more than it *actually* is, always on the way to another possibility of its own being. '*Dasein* is in each case already ahead of itself in its being. *Dasein* is always already "out beyond itself", not as a relating to other beings that it is *not*, but as being towards the potential for being that it itself is' [SZ 191–2], as Heidegger will put it in §42 of *Being and Time*. It is in precisely this sense that *Dasein*, in its being, is a *Seinkönnen* [SZ 144], an ability to be or a potentiality-for-being. Thus, as Heidegger argues in §31:

> *Dasein* is constantly 'more' than it factually is, supposing that one might want to make an inventory of it as something-at-hand and list the contents of its being, and supposing that one were able to do so. But it is never more than it factically is, for to its facticity potentiality-for-Being belongs

[26] Heidegger was certainly influenced by Søren Kierkegaard's transposition of Aristotle's modal categories within his religiously motivated psychological reflections, but another essay would be able to show that it is precisely through this interpretation of movement *as a mode of being* that the German philosopher goes beyond the remarkable and enigmatic reflections of the Dane.

[27] Heidegger, GA18 315/212. Husserl's account of the horizons constitutive of experience was significant for Heidegger's account of being and possibility, as we saw in the first section of this essay, and as Iain Macdonald contends in '"What is, is more than it is": Heidegger and Adorno and the Priority of Possibility', *International Journal of Philosophical Studies* 19/1 (2011): 31–57. It is, however, only an idea of movement that enables the break-through to an account of *Dasein*'s being as a being-possible; Husserl's static analysis of the ideal horizons constituting the present thing does not yet bring us to an idea of *Dasein* as stretched out beyond itself, beyond the present, according to the peculiar from-to structure characteristic of movement.

essentially. Yet as Being-possible, moreover, *Dasein* is never anything less; that is to say, it *is* existentially that which, in its potentiality-for-Being, it is *not yet*. [SZ 145][28]

The distinction between factuality and facticity is key here: if one thinks being 'factually' and, that is to say, according to a traditional idea of actuality, *Dasein*, in being stretched out and thus always ahead of itself, is either less or more than what it is. Yet the 'less' as much as the 'more' here presupposes an ontological standard that is wholly inadequate, Heidegger contends, to the *Faktizität* or facticity, to the particular kind of worldly and historical being of *Dasein* as a potentiality-for-being.

## 8.3 Temporality and the Modality of History

*Dasein*'s projective understanding of its own possibilities, which underlies its everyday commerce with the ready-to-hand things of the world, is, on Heidegger's account, the most fundamental form of our awareness of possibility; it is, as he puts it in 1928, 'the origin of possibility as such'.[29] That said, the claims of §31 of *Being and Time* concerning *Dasein* as a *Seinkönnen* can hardly be accepted, or even understood, without further development. The idea that *Dasein is* what it *is not yet* is problematic not least because it evidently contradicts a common, 'vulgar' conception of time as a succession of moments. If *Dasein* is what is not yet, it exists beyond the present and it is in some sense its future. That this is Heidegger's thought becomes explicit in part II of *Being and Time*, and particularly in §65 concerning 'The Temporality of the Understanding', where he begins to interpret the previous findings of the text in terms of temporality:

The projective self-understanding into an existential (*existenziellen*) possibility has for its ground the coming-to-itself from a given possibility, as which in each case *Dasein* exists. The future makes possible a being that is in such a way that it understandingly exists in its ability-to-be. The essentially futural projecting does not grasp the projected possibility thematically in a conceiving (*Meinen*), but rather projects itself into it as possibility. [SZ 336]

[28] Against the background of this paragraph, Judith Wolfe (*Heidegger's Eschatology: Theological Horizons in Martin Heidegger's Early Work*, Oxford: Oxford University Press, 2013, p. 119) argues that Heidegger's account of possibility in §31 of *Being and Time* 'can be criticised on Heidegger's own terms as a spatialization of being-as-possibility. Because *Dasein is* its possibility (in the present) rather than relating to any particular possibility (in the future), no particular choice or event actually matters for its essence'. On the contrary, every particular choice *Dasein* makes matters for it, because what it decides and does becomes—as we will see—its factical and thrown having-been which is constitutive of what, or better who, *Dasein* is. This analysis of possibility in part I of *Being and Time*, as will become clear below, does not contradict part II of the text, but leads to it. Wolfe does not elucidate why exactly Heidegger's analysis in these pages amounts to a spatialization of *Dasein*, and the additional contention that 'in thus "spatialising" possibility, Part I retains, despite its phenomenological method, characteristics of a *philosophia perennis*: a philosophy arrogating to itself a God's-eye-view from outside factic experience' is little more than an arbitrary assertion.

[29] Martin Heidegger, *Gestamtausgabe vol. 26: Metaphysische Anfangsgrunde der Logik im Ausgang von Leibniz*, edited by K. Held (Frankfurt am Main: Vittorio Klostermann, 1978), p. 244; *Metaphysical Foundations of Logic*, trans. M. Heim (Bloomington, IN: Indiana University Press, 1984), p. 189.

The being of *Dasein* consists in a pre-conceptual, projective understanding of its own individual and particular possibilities. But these *existenziell* possibilities are made possible by the structure of *Dasein*'s *Existenz* in such a way that *Dasein*, in 'projectively' understanding its possibilities, is always and already projected beyond itself. In this sense, the future makes possible the ability to be that is *Dasein*, and *Dasein* thus *is* its future. Moreover, insofar as it is the past that bequeaths 'given possibilities' to *Dasein*, Heidegger is also affirming that *Dasein* somehow is its past. The mode of being of *Dasein* somehow consists in being stretched out between its past and future; *Dasein is* this stretch.

It thus becomes clear that Heidegger's reflection on *Sein und Moglichkeit*, on being and possibility, leads to the critique of a 'vulgar' conception of time and the concomitant account of *Zeitlichkeit* or temporality presented in part II of *Sein und Zeit*. The vulgar conception of time is an expression of, to use contemporary philosophical terminology, 'presentism', in that it holds that the present, however fleeting it may be, is the only real aspect of time. Yet Heidegger is far from advocating 'eternalism' in opposition to 'presentism', since the former is equally an expression of a deep-rooted 'metaphysics of presence'. The eternalist merely extends the domain of the actual in holding that the past and future are equally as real or actual (i.e. present), as the present, the difference between these aspects of time being merely one of our limited perspective, of our frame of reference.[30] Heidegger's concern, in contrast, is to question the primacy of actuality and of presence, and their role as ontological standards. Being, he argues, at least in the case of *Dasein*'s being, involves the past and the future in such a way that being 'is' not simply presence. *Dasein* is its past, and is its future; and it is both in a way that goes beyond any ordinary or traditional conception of time reducing existence to the standards of the present. The past and future somehow exist—which is not to say that they are present or that they are in any sense things—because *Dasein*'s past and future are not a series of now-points that are, respectively, no longer or not yet present.

'Future [*Zukunft*]', in the most profound or original sense, 'does not mean a now that has *not yet* become actual and that sometime *will be* for the first time, but the coming [*Kunft*] in which *Dasein* comes toward itself in its ownmost potentiality for being' [SZ 325]. The future thus understood is so fundamental to *Dasein*'s being as an ability-to-be that Heidegger can claim its pre-eminence in relation to the past and present [SZ 337]. Nevertheless, the future is what it is only by means of the past, understood in the particular sense of the 'beenness' or *Gewesenheit* of *Dasein*; 'only insofar *Dasein* is as "I have been" can it futurally come toward itself in such a way that it comes back' [SZ 352]. The past in this sense is what it is by means of

[30] For some interesting remarks on Heidegger's account of time in relation to the presentism/eternalism distinction within 20th-century 'analytic' philosophy of time, see Jack Reynolds, 'The Analytic/Continental Divide: A *Contretemps*?' in *The Antipodean Philosopher*, Vol. 1, edited by G. Oppy, N. Rakakis, and L. Burns (Lanham, MD: Lexington, 2011), 239–54.

*Wiederhölung* [SZ 375], which is not merely a reiteration of the same, but repetition *with a difference*, a productive repetition that takes up what has been as a source of possibility for the future. *Dasein*'s having-been is not a realm of dead necessity, and yet the possibilities it bequeaths are what they are only in their repetition through the openness of the future. If the future presupposes the past, therefore, it does so only to the extent that having-been itself presupposes the future: 'Dasein can only be its been-ness insofar as it is futural' [SZ 352]. There is a mutual inherence of *Dasein*'s having-been and its futurity, and time as temporality in this sense 'does not mean a "succession" ('*Nacheinander*')' of what Heidegger terms the 'ecstases' of past, present, and future; 'the future is *not later* than having-been, and this is not *earlier* than the present' [SZ 350]. *Beenness* is not a series of 'nows' that are no longer, and the future is not a series of 'nows' that are not yet. Instead, the past, the present, and the future all 'occur', as it were, 'at the same time'.

If *Dasein*'s being is a form of movement or movedness, as Heidegger argues in part II of *Being and Time*, time is of the essence of this movedness. *Dasein* is the movedness of time, where original time is no simple 'passage', 'flow' or 'process', but 'ecstatic temporality' wherein the 'ecstases' of future, present, and past are not independent parts or mutually exclusive aspects of time.[31] According to this ecstatic temporal structure, it is not the case that possibilities bequeathed by *Dasein*'s having-been in any given situation simply pre-exist the present. They certainly do not pre-exist the present like the possible worlds that Leibniz's God surveys before the actualization of the best among them. They do not even pre-exist the present in the sense of constituting an ever-growing block of former actualities, an independently existing reservoir of possibilities into which *Dasein*, from time to time, can 'dip'. On Heidegger's account, possibilities are rather what they are only through their futural repetition, through their revitalizing retrieval, and do not exist independently of the latter.[32] Consequently, it makes little sense to wonder whether possibility chronologically precedes actuality or vice versa. Recall that in *Metaphysics* IX Aristotle is concerned to determine whether actuality is ontologically, epistemologically but also chronologically prior to possibility.[33] Once we recognize, however, with

[31] Joseph Carter ('Heidegger's *Sein zum Tode* as Radicalisation of Aristotle's Definition of *Kinesis*', p. 474) has asked: 'if temporality is the fundamental aspect of the being of *Dasein*, then why does Heidegger also remark that *Dasein* is constituted in terms of motion? Are these two ways at odds, or might there be something more to *Dasein's* temporality that is not made explicit in the text?' I hope to have clarified that what is not made explicit in *Being and Time* is how Heidegger's thinking—in the 1920s as a whole and also in the two published parts of the text—moves *from* an analysis of *Dasein*'s movedness as a potentiality-for-being *to* the account of temporality that that analysis requires and presupposes.

[32] In 1928 Heidegger explicitly criticizes Bergson's version of this 'growing-block' theory of the past in itself; see GA 26 266/206.

[33] Aristotle, *Metaphysics* IX, 1049b13. Aristotle's remarks concerning the chronological priority: 'possibility in one sense is prior, in another sense not'. In *Le problème de l'être chez Aristote* (Paris: Presses Universitaires de France, 1962, pp. 442–3, translation mine) Pierre Aubenque uses this dual response in support of his argument that Aristotle's account of movement is already ecstatic in Heidegger's sense: 'the debate concerning the respective priority of possibility and actuality is a false debate. Actuality and

Heidegger, that ecstatic temporality is ontologically prior to and makes possible *chronos* or clock-time, then we arrive at the insight that possibility is not *temporally* prior to actuality. Both arrive, as it were, 'at the same time', and, in the event, co-constitute the shock of the new.

*Dasein*, then, is stretched out ecstatically between its past and future in such a way that the possibilities bequeathed by its 'having-been' are intrinsically futural. It is only insofar as it is ecstatically stretched-out in this manner that possibility can be higher than actuality, and that *Dasein* can exist as a potentiality-for-being. As Heidegger puts it in a lecture course of 1928:

> ... in ourselves possibility is higher than actuality, because with *Dasein* itself this being-higher becomes existent. This being-higher [*Höhersein*] of the possible, vis-à-vis the actual, is existent only when temporality temporalizes itself [*sich zeitigt*].[34]

If *Dasein* were not 'ecstatically' projected beyond itself into its having-been and future, it could not be an ability-to-be, and would instead be something actually present—and, concomitantly, possibility could only be of a lower ontological status than actual presence conceived as the meaning of being. The 'original determinant of possibility, the origin of possibility itself', as Heidegger writes in 1927, is thus time insofar as it 'temporalizes itself'.[35]

The analysis of possibility and repetition in *Being and Time* applies not only to individual temporality but also to collective history. The possibilities bequeathed to *Dasein* come to it not only from its own individual past, but also, and perhaps primarily, from the past of its community, from the 'past of its generation' [SZ 20]. This past, as much as *Dasein*'s individual past, 'is not something which *follows along after it*, but something which already goes ahead of it' [SZ 20]. Moreover, Heidegger extends this analysis of possibility and repetition to our explicit knowledge of the past in the *study of* history.[36] He argues that the proper 'theme of *Historie*', of historical knowledge, 'is neither that which has happened just once for all nor something universal that floats above it, but the possibility which has been factically existent' [SZ 395]. A genuine or authentic mode of studying history should be concerned neither solely with recording past actualities in the sense of establishing what 'really happened', nor with attempting to discern necessary laws

possibility are co-originary; they are only the ecstases of movement; only the clash of possibility and actuality at the heart of movement is real; only the violence of human discourse... can maintain dissociated... the originary tension which constitutes, in its unity that is ever divided, the being of the being in movement.'

34 Heidegger, GA 26 280/216.

35 Heidegger, *Gestamtausgabe vol. 24: Die Grundprobleme der Phänomenologie*, edited by F.-W. von Herrmann (Frankfurt am Main: Klostermann, 1989), p. 463; *The Basic Problems of Phenomenology*, trans. A. Hofstadter (Bloomington, IN: Indiana University Press, 1982), p. 325.

36 It is not possible to engage here with the nature and legitimacy of Heidegger's move—if it is one—in *Being and Time* from an analysis of individual temporality to one of collective history.

governing the historical process. Instead, the study of history should be concerned primarily with possibility, and it will:

disclose the quiet force of the possible with greater penetration the more simply and the more concretely having-been-in-the-world is understood in terms of its possibility, and 'only' presented as such. [SZ 394]

The historical process is the history of human beings, each of which, as *Dasein*, is a being-possible and a genuine historical study has to take this potentiality-for-being as its primary object. One might say that historiology has to recognize human freedom as its proper object, and Heidegger could happily accept such a claim on condition that one follows his attempt in *Being and Time* to conceive freedom from the perspective of his analysis of possibility and ecstatic temporality. From this perspective, wondering whether Heidegger is proposing a mode of history either concerned or unconcerned with the 'facts' amounts to a false debate: if historical facts are understood merely as past actualities, then Heidegger certainly urges us to look beyond them, but if they are understood in their *facticity*, and that is to say, as a manifestation of *Dasein*'s potentiality-for-being, then the study of history, at least in the existentialist mode of history that Heidegger presupposes, should begin and end with them.[37]

Such an existentialist mode of historical study is grounded on a particular, authentic mode of *Dasein*'s being-historical, an authentic mode of *Geschichtlichkeit* or historicity, in which the past as possible is genuinely 'repeated' for the sake of the future: 'only by historicity [*Geschichtlichkeit*] which is factical and authentic can the history [*Historie*] of what has-been-there [ . . . ] be disclosed in such a manner that in repetition the "force" of the possible gets struck home into one's factical existence' [SZ 395].[38] In this sense, the historical world is the domain of the possible not simply because it is the history of former potentialities for being, but also because it is what it is only as a function of the futurity of *Dasein* in the present. We might ordinarily think that the historical past is a domain of necessity since we can no longer do

[37] Felix O'Murchadha responds to the claim of David Hoy ('History, Historicity and Historiography in *Being and Time*' in M. Murray (ed.), *Heidegger and Modern Philosophy*, New Haven: Yale University Press, 1978) that the historian should not be concerned with facts but with possibilities thus: '[i]t is indeed the case that Heidegger states the *theme* of historiography to be the possibility of having-been existence. The theme is, however, the "horizon" of a projection, which holds a particular region of entities [ . . . ]. Within this horizon are the objects of the specific sciences, the entities as present-at-hand. Historiography does not disclose its theme in its truth. It remains tied to its objects. Without understanding this difference between theme and object, Heidegger's attempt to transcend historiography and chronology must remain obscure' (*The Time of Revolution*, p. 27). Heidegger does not, however, attempt to 'transcend' historiography (i.e. the study of history) but rather to lead the historian to conduct it in the right way; and there is no justification for considering the 'theme' of historiography as a kind of transcendental condition that is presupposed by but not directly accessible to the historian.

[38] See Costantino Esposito, *Heidegger. Storia e fenomenologia del possibile* (Bari: Levante editori, 1992), and particularly its chapter 'La storiografia come scienza del possibile' for a longer exposition of Heidegger's claims concerning history and possibility.

anything about it, but, for Heidegger, the past in its sense and significance for us is still to come. Thus, as he writes in 1928:

> The actuality of what has been resides in its possibility. The possibility becomes manifest as the answer to a living question that sets before itself a futural present in the sense of 'what can we do?' The objectivity of the historical resides in the inexhaustibility of possibilities, and not in the fixed rigidity of a result.[39]

The study of history should not, *pace* the young Nietzsche of the second *Untimely Meditation*, be sometimes critical, sometimes antiquarian, and sometimes monumental. It should, as Heidegger contends in §76 of *Being and Time*, be all three all at once; and as always monumental, it should always be concerned with what we can do now and in the future, with the past as a source of possibility for the present and future.[40]

## 8.4 Death as the Possibility of Impossibility

No examination of Heidegger's treatment of possibility could hope to be comprehensive without discussion of his account of death (*Tod*) as the 'possibility of the absolute impossibility of *Dasein*' [SZ 262]. This account has long been a matter of controversy, but my aim here is to show how an adequate grasp of *Dasein*'s being-possible as movedness allows us better to understand it, and allows us to avoid the more extreme positions taken in relation to it within the secondary literature.

How can I understand death? I cannot experience *my own* death, given that I will no longer exist at the moment it occurs. I can experience only dying, in the sense of the moments before death, but not my death itself. 'Death', as Heidegger writes, 'gives *Dasein* nothing to be "actualized", nothing which *Dasein*, as actual, could itself *be*' [SZ 262]. For all that death is the end of my existence is not an actual event in that existence, not even the final one. Yet there is, Heidegger contends, another reason for death's lack of actuality: I do not experience death when witnessing somebody else 'pass away'. I certainly witness their passing from being-alive to being-dead, but I do not, despite our ordinary use of language and the gravity of the event, experience their death, for death, in its most proper sense, is always *my death*.[41] Death is, as Heidegger puts it, 'non-relational (*unbezüglich*)' [SZ 250], since no one can

[39] GA 26 88/72.

[40] Heidegger's relatively generous interpretation of Nietzsche's second *Untimely Meditation* in §76 of *Being and Time* is altered significantly in his 1938 seminar on the text; see *Gesamtausgabe vol. 46: Zur Auslegung von Nietszches II. Unzeitgemässer Betractung*, translation by U. Haase and M. Sinclair as *Interpretation of Nietzsche's 2nd Untimely Meditation* (Bloomington, IN: Indiana University Press, 2016). See also U. Haase and M. Sinclair, 'History and the Meaning of Life' in *Heidegger in the Twenty-First Century*, edited by T. Giorgakis and P. Ennis (Dordrecht: Springer, 2015).

[41] It is not possible here to address the numerous critiques of Heidegger's distinction between my death and the death of the other, but see Daniel Dahlstrom, 'Authenticity and the Absence of Death' in *Heidegger, Authenticity and the Self: Themes from Division Two of Being and Time*, edited by D. McManus (London: Routledge, 2015), 146–62 for a recent reflection on the issue.

experience my death with me, and in this sense one always dies alone. Death is the 'ownmost (*eigenste*)' [SZ 250] characteristic of *Dasein*, since no one can die in my place and it is radically individualizing. Someone can heroically save my life at the cost of his own, but nobody can take my death upon themselves in the sense of experiencing it for me, and nobody can save me from the necessity of facing it at some time.

Is death, then, a necessity? 'Nobody doubts that one dies' [SZ 257], and the fact that no exception has yet been found to the proposition that all men are mortal may seem to amount to some kind of necessity. Death is 'not to be outstripped (*unüberholbar*)' [SZ 251], and at some point, as we claim to know, my time will and must end. Yet this necessity is no logical necessity, and in the terms of an Aristotelian statistical or temporal account of modality this apparent necessity is merely a possibility; the possible on this account is what must be actualized at some point in time, in contrast to the necessary which is actual at all points in time. If death must occur at some point, then, according to this schema it resembles more a possibility than a necessity, even though my death can never be an actual event. Death resembles a possibility all the more in that it is 'indefinite as to its when' [SZ 258] and can happen at any time.

Heidegger's existential analysis of death, famously, attempts to account for death as a possibility, and the preceding remarks help us to understand why. Of course, the other possibilities of *Dasein* can be actualized—even though, as we have seen, *Dasein* can never simply be those possibilities once actualized for it is always on the way to another possibility of its own being—whereas *Dasein*'s death cannot. Yet this, Heidegger contends, takes nothing away from, and, in fact, only adds to, death's character as a possibility: death 'offers no support for becoming intent on something, "picturing" to oneself the actuality which is possible and so forgetting its possibility' [SZ 262]. That death is not the possibility of an actuality is, of course, one reason why Heidegger characterizes this possibility as the 'possibility of an impossibility'. Another reason is that this purported possibility amounts to *Dasein*'s no longer existing as a being-possible, to its no longer existing at all; '[i]ts death is the possibility of no-longer-being-able-to-be-there (*Nicht-mehr-dasein-könnens*)' [SZ 250].

For some, this talk of a possibility that is not the possibility of an actuality amounts to little more than empty speculation based on an abuse of language. As Paul Edwards has it, to 'describe the annihilation of all consciousness, the impossibility of every way of comporting oneself' as a possibility 'is carrying the misuse of language to the ultimate degree'.[42] Heidegger is and must be using the term 'possible' here in a particular sense, a sense that contrasts with the rest of *Being and Time* and that amounts to a non-sense. The term, Edwards argues, is redundant in the existential analysis of death, since Heidegger is speaking merely of the non-actuality, the total absence, of *Dasein* once

[42] Paul Edwards, 'Heidegger and Death as "Possibility"', *Mind* 84 (1975): 548–66, p. 558.

dead; and if he had described death simply as the 'impossibility of *Dasein*', as absolute non-existence, he would have produced much less confusion.

William Blattner has presented an influential response to such charges by defending Heidegger's use of 'possibility' as consistent with *Being and Time* as a whole, and by urging us to recognize instead that it is the term 'death' that does not have an ordinary sense in its existential analysis. Death does not mean the termination of *Dasein*'s 'life'—this is what Heidegger terms 'demise (*Ableben*)' [SZ 247]—but a particular way in which *Dasein* exists. As Heidegger writes: '[d]eath is a *way* to be, which *Dasein* takes up as soon as it is' [SZ 245]. For Blattner, this particular way of existing amounts to an anxiety attack: 'death is a condition in which *Dasein*'s being is at issue, but in which *Dasein* is anxiously unable to understand itself by projecting itself into some possible way to be'.[43] In such an anxiety attack, *Dasein* loses its grip on the world and all its particular possibilities appear equally meaningless. Death is thus a concrete possibility of existing, and I may have already 'died' several times; but this possibility is one whereby *Dasein*, in its anxiety, finds itself no longer able to project itself into any particular possibility—and thus no longer genuinely able to be. For Blattner, Heidegger's analysis thus presupposes two levels of existence: *Dasein*'s existence in a 'thin' sense, as disclosed in an anxiety attack, is distinct from, and the condition of, its existence in a 'thick' sense as projecting possibilities at any given time.

This interpretation certainly has the merit of stressing that Heidegger's analysis of death is an account of '*dying* [*Sterben*]' [SZ 247], which is a structure or aspect of its existence, and thus that death, existentially understood, is in some sense a phenomenon of life. Yet though death, for Heidegger, is not the termination of a life, it nevertheless remains the case that the structural features of death in this existential analysis are, as Iain Thomson has highlighted recently,[44] all borrowed from, and intrinsically related to, the life-terminating event that Heidegger names *demise*. Moreover, Heidegger does not describe dying as episodic in the way that Blattner's interpretation of it as an anxiety attack would require.[45] Finally, as Havi Carel has argued,[46] this interpretation removes Heidegger's analysis of death from the wider context of part II of *Being and Time*, from its concern for the temporality of *Dasein* and for the finitude of that temporality.

Is there, then, a way to defend Heidegger's account of death as the 'possibility of the impossibility of existence' from charges of obscurantism and redundancy that at the same time avoids the drawbacks of the decontexualizing interpretation proposed by Blattner? My contention here is that there is, and that this way relies on

[43] William Blattner, 'The Concept of Death in *Being and Time*', *Man and World* 27 (1994): 29–70.

[44] See Iain Thomson, 'Death and Demise in *Being and Time*'.

[45] For this critique of Blattner's interpretation, see Taylor Carman, 'Things Fall Apart' in *Heidegger, Authenticity and the Self: Themes from Division Two of Being and Time*, edited by D. McManus (London: Routledge, 2015), 135–45.

[46] See Havi Carel, 'Temporal Finitude and the Finitude of Possibility: the Double Meaning of Death in *Being and Time*', *International Journal of Philosophical Studies* 15/4 (2007): 541–56.

understanding adequately Heidegger's account of being-possible in terms of *Dasein*'s movedness.[47] 'Death', certainly, 'gives Dasein nothing to be "actualized", nothing which *Dasein*, as actual, could itself *be*' [SZ 262], but *Dasein* nevertheless *is* its death, at each and every moment of its life—*not as an actuality, but as a possibility. Dasein*, as we have seen, is the possibilities it is on the way to realizing, and those possibilities genuinely exist as possibilities only when it is on the way to realizing them. There is, however, one particular possibility that *Dasein* is on the way to realizing, and thus that it *is*, from the moment of its birth—and that possibility is its *death*. In short, Heidegger argues that *Dasein*, for as long as it is alive, *is* the non-actualizable possibility of its own death. There is no flat contradiction in the phrase the 'possibility of the impossibility' if we see that the 'possibility' constitutes *Dasein*'s being in the present, as a being-on-the-way to its death, whereas the 'impossibility' describes the non-actualizable end of *Dasein*'s existence that could befall it at any time, and to which *Dasein* is always and already heading. We do not need 'thick' and 'thin' notions of existence in order to see something other than a contradiction in Heidegger's formulation; we need simply to recognize that possibility, for Heidegger, is a *mode of being*. *Dasein is* its possibilities when it is on the way to realizing them, and there is one possibility that it always and already is. Certainly, the idea of a possibility that can never be realized is strange; but the strangeness of Heidegger's formulation attempts to describe the essential strangeness of the human condition; a condition in which, as Heidegger contends, the human being does have an individual and internal sense of its own mortality.

The term 'possibility' in Heidegger's existential analysis of death is, then, used in a sense consistent with the rest of *Being and Time*—but in order to understand this analysis we first have to understand the general sense of possibility in Heidegger's Aristotelian account of *Dasein*'s movedness. Of course, *Dasein*'s death as a possibility is distinct from all others. It is a non-contingent possibility that is independent of circumstance and the particular situation. Moreover, Heidegger is at great pains to show that the more *Dasein* understands and reveals itself as a finite, ever non-actualizable being-possible, the more it understands and reveals itself as mortal, with a limited-time span for any of the particular, realizable possibilities it endeavours to pursue: the 'anticipation' of death, as Heidegger has it, 'makes accessible *in* the possibility that cannot be outstripped all of the possibilities available for *Dasein*' [SZ 264]. Anticipation reveals possibilities as what they are, namely possibilities for and of a finite being-possible—and in this sense in death '*Dasein*'s character as possibility lets itself be revealed most precisely' [SZ 248].[48]

[47] My interpretation owes much to Joseph Carter's 'Heidegger's *Sein-zum-Tode* and Aristotle's Definition of Kinesis'. Carter does not, however, show how a proper understanding of *Dasein*'s movedness allows us to respond to Edwards' and Blattner's influential responses to Heidegger's analysis of death as a possibility.

[48] Concerning the specificity of death as a possibility, Steven Mulhall writes in 'Human Mortality: Heidegger on How to Portray the Impossible Possibility of *Dasein*' in *A Companion to Heidegger*, edited by

## 8.5 Possibility as Hidden Appropriateness

It remains to examine how Heidegger's conception of possibility undergoes a significant change in the 1930s—a change that, I contend, is pivotal in the turning or *Kehre* that marks his philosophical development in that decade. This change occurs by means of a reflection on the work of art, and through, more specifically, attention to art-production. In *Being and Time*, as we have seen, Heidegger had conceived possibility as a category according to an account of things in their use, according to the idea of *Zuhandenheit*. However, in and after 'The Origin of the Work of Art', an essay the third and final draft of which was written in 1935–36,[49] Heidegger reflects on the production of both artwork and equipment in a way that leads him to revise his conception of the modal status of the things that we are not.

'The Origin of the Work of Art' addresses the issue of art-production with reference to a dictum of the German Renaissance artist, Albrecht Dürer: 'in truth, art lies hidden within nature; he who can wrest (*reißen*) it from her, has it'. Heidegger takes this to mean that the finished form of the artwork lies dormant in nature, in the work materials, and that the artist has to coax the form from them. This entails, first of all, that the vision or knowing peculiar to creation is not to be understood as the envisaging, in abstraction from the work material, of an idea, or plan of the work to be realized, which idea could then be superimposed on, forced on that material. The vision consists much more in the capacity to apprehend what is possible for the material with which one is working; it consists in the capacity to apprehend the possibility of, for example, the statue in the stone, to apprehend what figure the stone itself is apt for or capable of. Art-production is less 'creative' than it is revelatory, and in realizing the design in the work material the artist does more, or less, and at any rate something other than act on an inert matter. In 'wresting' the figure from nature she rather *lets* the material come to presence in a definite figure, she brings this figure itself into presence, from a prior obscurity or state of hiddenness. Art-production,

Hubert L. Dreyfus and Mark A. Wrathall (Oxford: Blackwell, 2005), 297–310, p. 304: '[w]e cannot understand our relation to our own end on the model of our relation to any authentic possibility of our being—as if our death stood on the same level (the ontic or existentiell level) as any other possibility upon which we might project ourselves. Heidegger's point in calling our relation to our own end our "being-toward death" is to present it as an ontological (that is, existential) structure, rather than as one existentiell state (even a pervasive or common one) of the kind that that structure makes possible. In short, we cannot grasp Heidegger's account of death except against the horizon of his account of the ontological difference—the division between ontic and ontological matters.' It is not, however, possible to separate the 'existential' from the 'existentiell', or the ontic from the ontological in this way, for Heidegger's point is that any particular possibility is already ontological or existential in that *Dasein* is that possibility when it is on the way to realizing it. Iain Thomson is much closer to the truth when he notes that 'here as elsewhere, the ontic and the ontological are not *heterogeneous* domains (*pace* orthodox Heideggerians and influential critics like Habermas) but rather necessarily overlap and interpenetrate' ('Death and Demise in *Being and Time*', p. 278).

[49] For a genetic study of the development of Heidegger's thinking in the three versions of 'The Origin of the Work of Art', see the fifth chapter of my *Heidegger, Aristotle and the Work of Art.*

Heidegger contends, contains an essential passivity; it is 'a receiving and extracting [*Entnehmen*] within the relation to un-concealment'.[50]

Heidegger appeals to Dürer's dictum in 'The Origin of the Work of Art' in order to distinguish what we traditionally call 'fine art' from the merely mechanical arts. The 'fine' artist clearly does not produce something that will recede from our attention to the degree that it is successfully used, but rather makes something that will stand *before the eyes* (*or ears*), often in a peculiar isolation from all other things; but, as Heidegger also contends, the particular form of the artwork is drawn from rather than imposed in a mechanical fashion on what we call the 'work-material'.[51] Yet Heidegger's strategy is interrupted by the fact that Dürer's dictum expresses an idea that finds its origin in the philosophy of Aristotle, who makes no such distinction between 'fine art' and craft production. Dürer's dictum expresses a commonplace of Renaissance art-theory[52]—the idea that the work is hidden in the work material before it is unearthed in the process of production—but this commonplace ultimately derives from Aristotle's account of possibility as potentiality. In *Metaphysics* IX, Aristotle distinguishes actuality from possibility thus:

> *energeia* means the presence [*to huparchein*] of the thing but not in the sense which we mean by *potentiality* [*dunamei*]. We say that a thing is present potentially as Hermes is present in the wood.[53]

Although the example is of a statue, the idea of potentiality here applies to craft production or *poiesis* in general. In production as such the product is potentially present in the work material—and since the wood *is* the statue potentially, the latter needs only to be wrought out, as Aristotle continues, from the former by a process of *aphairesis*, a process of abstraction.[54]

One might understand possibility in this sense as a state of in-determination: before the actualization of the specific form of the statue, the wood is in a mere state of in-determination, a state from which other forms or determinations could have emerged than those in fact actualized. In 1939, however, within a reading of Book II, chapter II of Aristotle's *Physics,* Heidegger interprets this sense of *dunamis* more positively in translating it as 'appropriate-for [*Eignung-zu*]':

> 'Appropriate-for' means: tailored to the appearance of a table, hence for that wherein the generating of the table—the movement (*metabole*)—comes to its *end.* The change of the appropriate wood into a table consists in the fact that the very appropriateness of what is

[50] 'Der Ursprung des Kunstwerkes' in *Gesamtausgabe vol. 5*: *Holzwege*, edited by F.-W. von Herrmann (Frankfurt am Main: Klostermann, 1994), 1–74, p. 50; 'The Origin of the Work of Art' in *Off the Beaten Track*, trans. J. Young and K. Haynes (Cambridge: Cambridge University Press, 2002), 1–56, p. 37.

[51] GA5 34/25. [52] See Erwin Panofsky, *Idea*, trans. J. S. Peake (New York: Harper Collins, 1975).

[53] Aristotle, *Metaphysics,* 1048a32–3.

[54] On the manifold meanings of *aphairesis* in Aristotle, see Joseph Owens, *The Doctrine of Being in the Aristotelian Metaphysics* (Toronto: Pontifical Institute of Medieval Studies, 1978), pp. 382–5, and John J. Cleary, 'On the Terminology of "Abstraction" in Aristotle', *Phronesis* 30/1 (1985): 13–45.

appropriated emerges more fully into view and reaches its fulfilment in the appearance of a table and thus comes to stand in the table that has been produced, placed *forth*, i.e. into the unhidden.[55]

Wood is appropriate for the making of a table, and this appropriateness comes into view all the more clearly, as we have seen, in the process of production. The appropriateness of the wood for the table is what is appropriated by the producer in the process of production. To be appropriated in this sense, however, means to be brought forth from a prior state of invisible latency into the light, into presence. Heidegger reads in Aristotle's thinking, then, a contrast between a latent, hidden capacity or appropriateness and its bringing to light, between a hidden capacity and its revelation, which revelation is an emergence into the 'unhidden'. Certainly, Aristotle might be taken to speak simply of a difference between in-determination and determination, between that which is formed, and that which is, relatively speaking, formless. Yet this, for Heidegger, would be to fail to grasp adequately what it means for Aristotle to hold that the form is already present, but hidden, in the work-material, and thus that the form needs only to be 'abstracted' from that material.

In returning to Aristotle's *Metaphysics*, then, Heidegger comes to see the possibility of an element of revelation in all modes of manual production, in both 'fine art' and 'craft'. In so doing, he alters the interpretation of *dunamis* that he had advanced in the 1924 *Basic Concepts of Aristotelian Philosophy*. Here he had interpreted *dunamis* in terms that point to the analysis of *Zuhandenheit* in *Being and Time*, even though the focus is on the utility of the work-material from which the product is to be produced, rather than on the utility of the finished product:

The tree-trunk can present itself to me according to its character of serviceability for (*Dienbarkeit zu*), its availability for boat-building. This tree-trunk has the character of being-serviceable (*Dienlichsein*) for, of usability (*Verwendbarkeit*) for [...], not because I apprehend it in this way, but rather it is its way of being. [...] *Dunamei*-being is a *positive determination of the way of its there*. For a long time I have preferred to call this being-character of things significance [*Bedeutsamkeit*]. This character of being is the primary one in which the world shows itself to us.[56]

In the 1930s, this early reading of *dunamis* is not negated but enlarged: *dunamis* certainly involves an idea of being-appropriate for something else, and thus a certain idea of utility. Yet appropriateness is, Heidegger contends, a *hidden* appropriateness, an ability to come into presence—to, for, and before the carpenter working on the

[55] Heidegger, *Gesamtausgabe vol. 9: Wegmarken*, edited by F.-W. von Herrmann (Frankfurt am Main: Klostermann, 1976), p. 350; *Pathmarks*, edited by W. McNeill (Cambridge: Cambridge University Press, 1998), p. 214.

[56] GA18 300/134. This reading of Aristotle is echoed in *Being and Time* when Heidegger talks of natural things as natural resources: 'The wood is a forest of timber, the mountain a quarry of rock; the river is water-power, the wind is wind "in the sails"' [SZ 70]. Certainly, by 1927 Heidegger has curtailed the generosity of his reading of Aristotle: 'the specifically "pragmatic" character of the *pragmata* is just what the Greeks left in obscurity' [SZ 68], and no more will he attempt to retrieve a concept of *Sein-in-einer-Welt* from Aristotle.

tree-trunk—that is irreducible to any conception of utility, to the purposes of the agent, even if these purposes are thought ontologically according to the idea of *Zuhandenheit*. The idea of *Zuhandenheit* passes over the *hidden* appropriateness of the work-material, and the ability to enter into the unhidden that constitutes its very being.

This change in interpretation of Aristotle's concept of *dunamis* may appear to amount to little more than a nuance.[57] Yet everything in Heidegger's changed interpretation of Aristotle in the 1930s turns on this apparent nuance: this account of *dunamis* enables Heidegger to claim that in Greek thinking matter (*hule*) is not simply inert matter, just as the producer is not an *efficient cause* forcing the wood to become simply what it is not. At the same time, it allows him to argue that the master-word of Aristotle's metaphysics, namely *ousia*—which is ordinarily translated as being—should be understood as:

> *Anwesung*, presencing, instead of *Anwesenheit*, presence. What we mean is not *Vorhandenheit*, and certainly not something that is exhausted in mere stability; rather: presencing, in the sense of coming forth into the unhidden, placing itself into the open. One does not get at the meaning of presencing by referring to mere duration.[58]

Finally, it allows him to argue that Aristotle's conception of *energeia*, understood verbally as a having-been-released into presence, is fundamentally different to the Latin *actus*; and to argue that with this Latin translation of *energeia*, with one fell swoop 'the Greek world', i.e. the Greek understanding of being, 'was toppled'.[59]

We should note also that this changed conception of production leads Heidegger in 'The Origin of the Work of Art' to reconsider the being—the 'modality'—of the finished product beyond the idea of *Zuhandenheit*. He selects as an example, not equipment held in the hand, but equipment worn on the feet, a pair of boots belonging to a peasant-woman. In the field, before the peasant might take them to the cobbler when they require repair, the shoes have, Heidegger contends, a *reliability* which is prior not only to their *presence before the eyes* or *Vorhandenheit* but also to their *Zuhandenheit*. It is because of, by virtue of the reliability of a pair of shoes—if they are in good state of repair—and only by virtue of this reliability that the peasant can have particular projects to pursue: 'the equipmental being of the equipment consists indeed in its usefulness. But this usefulness rests in the abundance of an essential Being of the equipment. We call it reliability (*Verlässlichkeit*)'.[60] The reliability of equipment is the prior condition of its utility and the 'latter vibrates in the former'; it 'would be nothing without it' and is its 'essential consequence'.[61] The peasant takes them for granted, but what is granted in this taking-for-granted is a

[57] The importance of this nuance has consistently been overlooked by Thomas Sheehan in his once ground-breaking work on Heidegger's interpretation of Aristotle. See, for example, 'On the Way to *Ereignis*: Heidegger's Interpretation of *Phusis*' in *Continental Philosophy in America* (Pittsburgh: Duquesne University Press, 1983), 131–64.

[58] GA9 272/208. [59] GA9 286/218. [60] GA5 19–20/14. [61] GA5 20/15.

form of life that precedes her explicit projects and even any form of purposiveness. The shoes open possibilities for the peasant, it might be said, but these possibilities are prior to any purposiveness. To be sure, this thesis is phenomenologically problematic, for the reliability that Heidegger attempts to expose does not manifest itself directly in experience. It does not even appear in the breakdown of the item of equipment, as *Being and Time* had argued concerning *Zuhandenheit.*[62] This, it would seem, is the principal reason why Heidegger introduces it, (in)famously, by way of an interpretation of a Van Gogh painting. Yet in tracing how it follows from his changed interpretation of production and *dunamis*, and thus from his recognition that the truth of what is transcends any mere utility, we are at least in a position to understand why he advances the thesis.

## 8.6 Conclusion: Being as Possibility in the Later Heidegger

In the mid-1930s, then, a significant change occurs in Heidegger's conception of possibility as a category, i.e. as an ontological determination of the things that we are not. Understanding this takes us some of the way to grasping the particular sense in which Heidegger, shortly after his reflection on art, in the *Contributions to Philosophy,* ponders 'another beginning' in philosophy that would set itself the task of thinking being as the possible. As we have seen, Heidegger contends that 'the *possible* (*das Mögliche*) essentially occurs in being [*Seyn*] alone and as its deepest fissure, so that in the thinking of the other beginning being must first be thought in the form of the possible'.[63] To think of being as the possible, is to think of it—not simply in terms of *Dasein*'s movedness, and beyond the idea of *Zuhandenheit*—as a hidden appropriateness first of all, and thus as an element able to grant or bestow a particular configuration of presence to, for and before the human being. Yet from the *Contributions to Philosophy* onwards, Heidegger attempts to radicalize and generalize this (neo)-Aristotelian insight concerning hidden appropriateness in craft production: being is a bestowal or granting, not simply or solely insofar as it characterizes the peculiar form of 'appropriation' that is manual production, but rather in that it constitutes the appearance of beings as such, the appearing of beings to and for the human being. Beings are granted by being, and being as such a granting is precisely what is to be thought as the possible:

> that being [*Sein*] is, and therefore does not become a being—this can be expressed most pointedly by saying that be-ing [*Seyn*] is possibility, something that is never objectively present and yet is always bestowing and denying itself in refusal through ap-propriation (*Er-eignis*).[64]

[62] On this point, see my *Heidegger, Aristotle and the Work of Art*, p. 154.
[63] Heidegger, GA 65: 475.
[64] Heidegger, GA 65: 475.

Being as such is to be conceived verbally as a happening or *event*—the ordinary meaning of the German *Ereignis*—, an 'event of appropriation' whereby, in our intentional and purposeful comportment towards ourselves and other things, beings are granted and bestowed by being. Being 'is' the coming into presence, the coming into the open of beings—and if we can hear in the word possible something other than a mere state or a static transcendental condition, then an idea of possibility can serve as a guide in our attempts to grasp this. Yet it is crucial, for Heidegger, to recognize adequately—in a way that, he claims, the thinkers of the 'first beginning' of philosophy in ancient Greece did not—that being as presencing does not deliver over its own secrets, and is as much a refusal and denial as it is a donation or granting—for being, of course, is never present and available to us as a being. Thus if being is the possible, it is, as it were, a 'possibilizing' or enabling of beings that, as an inexhaustible capacity, maintains its own reserve.

This allows us to approach, finally, Heidegger's remarks in the 1946 'Letter on Humanism' concerning being as possibility. Here he underlines once again that possibility in his sense must be distinguished from any notion of possibility as subordinate to actuality:

> Our words *möglich* and *Möglichkeit*, under the dominance of 'logic' and 'metaphysics', are thought solely in contrast to 'actuality'; that is, they are thought on the basis of a definite—the metaphysical—interpretation of being as *actus* and *potentia*... When I speak of the 'quiet power of the possible' I do not mean the *possibile* of a merely represented *possibilitas*, nor *potentia* as the *essentia* of an *actus* of *existentia;* rather, I mean being itself, which in its favouring [*mögend*] enables [*vermag*] thinking and hence the essence of humanity, and that means its relation to being. To enable [*vermögen*] something here means to preserve it in its essence, to maintain it in its element.[65]

Being, thought as the possible, and thought verbally as an enabling coming to presence, grants beings to the human being and enables what, for Heidegger, is proper to it, namely thinking:

> This enabling is what is properly 'possible' [*das Mögliche*], whose essence resides in favouring. From this favouring being enables thinking. The former makes the latter possible. Being is the enabling-favouring, the 'may be' [*das Mög-liche*]. As the element, being is the 'quiet power' of the favouring-enabling, that is, of the possible.[66]

Heidegger draws, then, on the verbal root of the German word for possibility: *Möglichkeit*, possibility, is a function of a certain *mögen*, a liking, granting or favouring that is, he contends, the 'essence' of being itself. That he now claims that this favouring enables thinking amounts to the recognition that *Dasein,* as a being-possible and in its ecstatic temporality, is itself bestowed, favoured, and enabled by being as such. As he will put it in the later essay *Time and Being*—an essay which obviously refers back

[65] GA9 316–17/242. [66] GA9 316/242.

to his master work of 1927—when grasped verbally as a presencing, being is akin to a fourth dimension of time: what unifies and first grants the ecstatic unity of future, present, and past is the entry into being of this unity itself, the favouring or enabling of this ecstatic unity by being.[67] This by no means constitutes a *volte face* in relation to *Being and Time*, but only a decentring of *Dasein*'s understanding of being in Heidegger's philosophy, and a new prioritization of that which it understands, namely being, understood as an enabling coming-to-presence—as the 'possible'—that nevertheless maintains its own absence.

## Bibliography

Aubenque, P., *Le problème de l'être chez Aristote* (Paris: Presses Universitaires de France, 1962).

Bergson, *La pensée et le mouvant*, edited by A. Bouaniche et al (Paris: Presses Universitaires de France, 2009).

Blattner, W., 'The Concept of Death in *Being and Time*', *Man and World* 27 (1994): 29–70.

Carel, H., 'Temporal Finitude and the Finitude of Possibility: the Double Meaning of Death in *Being and Time*', *International Journal of Philosophical Studies* 15/4 (2007): 541–56.

Carman, T., 'Things Fall Apart' in *Heidegger, Authenticity and the Self: Themes from Division Two of Being and Time*, edited by D. McManus (London: Routledge, 2015), 135–45.

Carter, J., 'Heidegger's *Sein zum Tode* as Radicalization of Aristotle's Definition of *Kinesis*', *Epoché* 18/2 (2014): 473–502.

Cleary, J., 'On the Terminology of 'Abstraction' in Aristotle', *Phronesis* 30/1 (1985): 13–45.

Dahlstrom, D., 'Authenticity and the Absence of Death' in *Heidegger, Authenticity and the Self: Themes from Division Two of Being and Time*, edited by D. McManus (London: Routledge, 2015).

Edwards, P., 'Heidegger and Death as "Possibility" ', *Mind* 84 (1975): 548–66.

Esposito, C., *Heidegger. Storia e fenomenologia del possibile* (Bari: Levante editori, 1992).

Gonzalez, F., 'Whose Metaphysics of Presence? Heidegger's Interpretation of *Energeia and Dunamis*', *The Southern Journal of Philosophy* 44 (2006): 533–68.

Haase, U. and Sinclair, M., 'History and the Meaning of Life' in *Heidegger in the Twenty-First Century*, edited by T. Giorgakis and P. Ennis (Dordrecht: Springer, 2015).

Han-Pile, B., 'Freedom and the "Choice to Choose to Oneself"' in *The Cambridge Companion to Being and Time*, edited by M. Wrathall (Cambridge: Cambridge University Press, 2013), 291–319.

Heidegger, M., *Gesamtausgabe vol. 9: Wegmarken*, edited by F.-W. von Herrmann (Frankfurt am Main: Klostermann, 1976); *Pathmarks*, edited by W. McNeill (Cambridge: Cambridge University Press, 1998).

Heidegger, M., *Zur Sache des Denkens* (Tübingen: Niemeyer, 1976); *On Time and Being*, trans. J. Stambaugh (New York: Harper & Row, 1972).

[67] See Heidegger, *Zur Sache des Denkens* (Tübingen: Niemeyer, 1976), p. 16; *On Time and Being*, tr. J. Stambaugh (New York: Harper & Row, 1972), p. 15.

Heidegger, M., *Gestamtausgabe vol. 26: Metaphysische Anfangsgrunde der Logik im Ausgang von Leibniz*, edited by K. Held (Frankfurt am Main: Vittorio Klostermann, 1978); *Metaphysical Foundations of Logic*, trans. M. Heim (Bloomington, IN: Indiana University Press, 1984).

Heidegger, M., *Sein und Zeit* (Tübingen: Max Niemeyer, 1984).

Heidegger, M., *Gestamtausgabe vol. 24: Die Grundprobleme der Phänomenologie*, edited by F.-W. von Herrmann (Frankfurt am Main: Klostermann, 1989); *The Basic Problems of Phenomenology*, trans. A. Hofstadter (Bloomington, IN: Indiana University Press, 1982).

Heidegger, M., *Gesamtausgabe* vol. 5: *Holzwege*, edited by F. W. von Herrmann (Frankfurt am Main: Klostermann, 1994); *Off the Beaten Track*, trans. J. Young and K. Haynes (Cambridge: Cambridge University Press, 2002).

Heidegger, M., *Gesamtausgabe vol. 65: Beiträge zur Philosophie (Vom Ereignis)* (Frankfurt am Main: Vittorio Klostermann, 1994); *Contributions to Philosophy (Of the Event)*, trans. R. Rojecwicz and D. Vallega-Neu (Bloomington, IN: Indiana University Press, 2012).

Heidegger, M., *Gesamtausgae vol. 21: Logik: Die Frage nach der Wahrheit*, edited by W. Biemel (Frankfurt am Main: Klostermann, 1995); *Logic: The Question of Truth*, trans. T. Sheehan (Bloomington, IN: Indiana University Press, 2010).

Heidegger, M., *Gesamtausgabe vol. 18: Grundbegriffe der aristotelischen Philosophie* (Frankfurt am Main: Vittorio Klostermann, 2002); *The Basic Concepts of Aristotelian Philosophy*, trans. R. Metcalfe and M. Tanzer (Bloomington, IN: Indiana University Press, 2009).

Heidegger, M., *Gesamtausgabe vol. 46: Zur Auslegung von Nietszches II. Unzeitgemässer Betractung*, edited by H. J. Friedrich (Frankfurt am Main: Vittorio Klostermann, 2003).

Hoy, D., 'History, Historicity and Historiography in *Being and Time*' in *Heidegger and Modern Philosophy*, edited by M. Murray (New Haven, CT: Yale University Press, 1978), 329–53.

Husserl, E., *Cartesian Meditations* (The Hague: Martinus Nijhoff, 1960).

Inwood, M., *A Heidegger Dictionary* (Oxford: Blackwell, 1999).

Kearney, R., 'Heidegger, The Possible and God' in *Heidegger et la question du dieu*, edited by R. Kearney and J. O'Leary (Paris: Grasset, 1981); republished in Heidegger, *Critical Assessments* vol. 4, edited by C. McCann (London: Routledge, 1992), 299–324.

Kearney, R., *La poétique du possible: phénoménologie herméneutique de la figuration* (Paris: Beauchesne, 1984).

Kosman, A., 'Aristotle's Definition of Motion', *Phronesis* 14/1 (1969): 40–62.

Kosman, A., *The Activity of Being: An Essay on Aristotle's Ontology* (Cambridge, MA: Harvard University Press, 2013).

Macdonald, I., '"What is, is more than it is": Heidegger and Adorno and the Priority of Possibility', *International Journal of Philosophical Studies* 19/1 (2011): 31–57.

McNeill, W., 'Rethinking the Possible: On the Radicalisation of Possibility in Heidegger's *Being and Time*', *theory@buffalo* 13 (2009): 105–125.

Mohanty, J. N., 'Husserl on "Possibility"', *Husserl Studies* 1 (1984): 13–29.

Mulhall, S., 'Human Mortality: Heidegger on How to Portray the Impossible Possibility of *Dasein*' in *A Companion to Heidegger*, edited by Hubert L. Dreyfus and Mark A. Wrathall (Oxford: Blackwell, 2005), 297–310.

Müller-Lauter, W., *Möglichkeit und Wirklichkeit bei Martin Heidegger* (De Gruyter: Berlin, 1960).

O'Murchadha, F., *The Time of Revolution: Kairos and Chronos in Heidegger* (London: Continuum, 2013).

Owens, J., *The Doctrine of Being in the Aristotelian Metaphysics* (Toronto: Pontifical Institute of Medieval Studies, 1978).

Panofsky, R., *Idea*, trans. J. S. Peake (New York: Harper Collins, 1975).

Reynolds, J., 'The Analytic/Continental Divide: A *Contretemps*?' in *The Antipodean Philosopher*, vol. 1, edited by G. Oppy, N. Rakakis, and L. Burns (Lanham, MD: Lexington, 2011), 239–54.

Sheehan, T., 'On the Way to *Ereignis*: Heidegger's Interpretation of *Phusis*' in *Continental Philosophy in America* (Pittsburgh, PA: Duquesne University Press, 1983), 131–64.

Sinclair, M., *Heidegger, Aristotle and the Work of Art* (Basingstoke: Palgrave, 2006).

Sinclair, M., 'Bergson on Possibility and Novelty', *Archiv für Geschichte der Philosophie* 96/1 (2014): 104–25.

Thomson, I., 'Death and Demise in *Being and Time*', *The Cambridge Companion to Being and Time*, edited by M. Wrathall (Cambridge: Cambridge University Press, 2013), 260–90.

Wolfe, J., *Heidegger's Eschatology: Theological Horizons in Martin Heidegger's Early Work* (Oxford: Oxford University Press, 2013).

# 9

# De Re Modality in the Late Twentieth Century

## The Prescient Quine

*John Divers*

Quine's (in)famous sceptical critique of de re modality is expounded in the pair of 1953 classic papers 'Reference and Modality'[1] and 'Three Grades of Modal Involvement'.[2] Here, I position the salient, and non-sceptical, treatments of de re modality in the later part of the twentieth century—those due to Kripke, Lewis, and Fine—in relation to that prior sceptical critique. I emphasize the insights on which Quine's scepticism was based and commend these as sound and enduring.[3]

## 9.1 Quine's Scepticism

In *Three Grades of Modal Involvement* Quine (pp. 158–9) locates our subject matter within a systematic and philosophically neutral scheme of logical syntax. De re modal predication is what makes for Grade 3 in that scheme, and to explain what that amounts to we proceed, with Quine, via Grades 1 and 2.[4]

[1] W. V. O. Quine, 'Reference and Modality', in *From a Logical Point of View* (Cambridge, MA: Harvard University Press, 1953). Page references are to the second edition of the text (New York: Harper and Row, 1961, 139–59), and henceforth I refer to it with the abbreviation RM.

[2] Quine, 'Three Grades of Modal Involvement', *Proceedings of the XIth International Congress of Philosophy*, Volume 14 (Amsterdam: North-Holland Publishing Co., 1953). Page references are to the reprint in Quine, *The Ways of Paradox and Other Essays,* revised edition (Cambridge, MA: Harvard University Press, 1976), 158–76, and henceforth I refer to this essay with the abbreviation TGMI.

[3] By inviting greater appreciation for Quine's achievements in this regard, I join J. Burgess, 'Quinusab omni naevo vindicatus' in *Mathematics, Models and Modality* (Cambridge: Cambridge University Press, 2008), 203–35.

[4] It would seem that Quine's method and starting point here is attributable to his Carnapian education: in particular, the approach to modal words as quasi-syntactic expressions that lead to metaphysical entanglements that we must attempt to avoid, wherever possible, by translation into explicit syntactical predicates. For a most informative discussion of Carnap's influence on Quine's modal scepticism in this respect, and in others, see Shieh, 'Modality' in *The Oxford Handbook of the History of Analytic Philosphy*, edited by M. Beaney (Oxford: Oxford University Press, 2013), esp. §36.4–5.

Grade 1 modality is intentionally and explicitly metalinguistic. Here, a modal predicate (of necessity) attaches to the name of a (closed) sentence and so says something of, or about, that sentence—thus:

(G1) *'Everything is physical'* **is necessary**.

Grade 2 modality by contrast, is located at the same syntactic level as its non-modal complement. Now, modality no longer stands above the object-language alongside the other means we have of talking about parts of that object-language, such as its sentences and words. Modality is now part of the object-language and it is part of our means of talking in that language about things other than languages. The modal expressions of Grade 2 are operators (*It is necessary that* __) of the kind that operate on closed, statement-expressing sentences (*everything is physical*) to make more complex closed, statement-expressing sentences—thus:

(G2) *It is necessary that everything is physical.*

In sentential modal logic (also known as propositional modal logic) this kind of sentence is regimented as $\Box$A and the iteration of modal operators, $\Box\Box A$ (*It is necessary that it is necessary that everything is physical*) is syntactically permitted. In that respect, the Grade 2 modal expressions function like other familiar operators such as the negation operator (*It is not the case that*...). So syntactically, the difference between Grade 1 and Grade 2 modality replicates the difference between (F1) and (F2):

(F1) 'Everything is physical' is not so.

(F2) It is not the case that everything is physical.

In both the cases of modality and negation, it is a good question why we have both sorts of construction (and how they relate to each other) but is a question beyond our immediate concerns with logical syntax. Continuing with those syntactic concerns, we might choose to open up the structure of the non-modal sentences of sentential modal logic to display their quantificational structure—thus, for the representative of the structure of *everything is physical* we would naturally have $\forall xPx$. But it still counts only as Grade 2 of modal involvement if the modal operators only have immediately within their scope, closed (statement-expressing) sentences—thus:

(G2*) $\Box\forall xPx$

In (G2*), we are certainly using the characteristic symbols of quantified modal logic, in which modal operators combine with quantifiers, variables and predicates, to regiment (G2). But the move to Grade 3, and to quantified modal logic proper, comes only when we allow as well-formed formulas a further kind of combination of these symbols. That is where a modal operator $\Box$ operates immediately on a predicate $Px$ (*x is physical*) to make a more complex modal predicate $\Box Px$, (*x is necessarily physical*): and, just to simplify the statement of syntax rules, all such predicates are also called 'open sentences'. Grade 3, then, allows as well-formed those formulas that we get when we close these open-sentences (predicates) to make closed sentences (statement-expressors) by placing some appropriate quantifier on the outside—thus:

(G3) Everything is necessarily physical

(G3*) $\forall x \Box Px$

So the definitive Grade 3 phenomenon is the occurrence of a modal operator (somewhere) between a quantifier-plus-variable and the variable that it binds: it is quantification *across*, past or beyond a modal operator.[5] If we apply the notion of *de re modal predication* to the language of quantified modal logic it is (the appearance within sentences of) predication of exactly that kind. In such de re modal predication, a modal operator (of necessity, for example) contributes to the formation of a modal expression which is apt to predicate something modal of whatever the values of the variables are: of the things that the language is interpreted as being about. That much ought to be philosophically uncontroversial, and so ought the following statement: it is the purpose of quantified modal logic(s) to treat the inferential properties of modal expressions at Grade 3 (and encompassing those at Grade 2 as a special case).

Against this background, Quine takes as methodologically equivalent commitment to the intelligibility of de re modal predication and commitment to the (semantic) adequacy of quantified modal logics. Quine's scepticism about de re modality is precisely the view that these twin commitments ought to be refused. What remains to be understood, then, is the case that Quine makes for his sceptical refusal of the de re modal package. In headline, Quine's case is (quite predictably) that we are to refuse the package because the associated benefits cannot be purchased at acceptable costs. In the remainder of this paper, and reflecting the approach of Quine, I shall have nothing at all to say about what such benefits might be. Thus, I shall concentrate entirely on what Quine characterizes as the costs of accepting de re modality.

## 9.2 Quine's Case for Scepticism

In my understanding, the master-argument of 'Reference and Modality' is as follows. Grade 3 modal contexts are prima facie referentially opaque [RM §1, 139–44]. Since no one can tolerate unexpurgated referential opacity in these contexts [RM §2, 144–50], the only questions are whether, and at what costs, Grade 3 modal contexts can effectively be purged of it. Then we have a dilemma. There are two broad strategies for purging referential opacity: the first of these (the language-dependence strategy) is demonstrably ineffective [RM §3, 150–4], while the second (the language-independence strategy) can be implemented only at unacceptable costs [RM §3, 154–6].

[5] The standard locution is 'quantifying into' a modal context. But, if left as that, this is a badly misleading description of the intended Grade 3 syntactic phenomenon. For, in my hearing at least, 'quantifying in' permits the understanding that a quantifier be put *inside* the modal operator. And that, being a Grade 2 construction, is exactly what is *not* intended.

The demonstration of the prima facie referential opacity of de re modal predication comes in the famous *number of planets* paradox:

(N1) The number of planets is 9

(N2) 9 is necessarily greater than 7

---

(N3) The number of planets is necessarily greater than 7.

We have a paradox in that it *appears* that all of the following conditions obtain: (N1) is true, (N2) is true, (N3) is false and the step that takes us from the premises, (N1) and (N2), to the conclusion, (N3), is an application of an impeccable inferential principle (viz. the inter-substitutivity of identicals).[6] Two points about the subsequent tightening-up of the case for referential opacity are to be noted.

Firstly, Quine feels bound to put his observation on a firmer footing by showing that an appropriate analogue of the number-of-planets paradox can be constructed at a more secure and significant level of syntax. So Quine attempts to free the issue from incidental considerations about the behaviour and introduction of singular terms, either in English or in quantificational logic. He does so by drilling down to the level of pure quantificational modal logic at which the only vocabulary is as follows: non-modal predicates, modal operators, sentence connectives, variables, and the quantifiers that bind those variables. And what Quine finds here is the prospect of a failure of the inter-substitutivity of identicals that is utterly inescapable—inescapable because (contrast modal English) there is no deeper level at which it might be analysed away—thus:

(N1*) $\Box Fx$

(N2*) $x=y$

---

(N3*) $\sim\Box Fy$

Quine's negotiation of this syntactic transition, from our N-version of the paradox to our N*-version of the paradox, is something that has been much discussed.[7] But here I settle simply for asserting that, for our purposes, these details can be bypassed.

Secondly, Quine is perfectly clear that, at this stage in the dialectic, the necessity (modality) involved is of a kind that Carnap and others in the broad camp of logical empiricism had been prepared to champion:[8] that is, analytic necessity, or analytic-or-logical necessity or broadly logical necessity. Following usage established in the early twentieth century Quine tends to call this *strict* necessity [e.g. RM 143]. And he further refines his target by contrasting this strict necessity with the most prominent case of non-strict necessities: that is the physical or causal modalities that feature

[6] See RM §1 and TGMI §I. Perhaps the 'apparent truth' of (N1) requires trans-generation explanation. Pluto was then classified as a planet and that the planets in question are supposed to be those in our solar system.

[7] On Quine's treatment of the related issues see Fine, 'Quine on Quantifying In' in *Modality and Tense: Philosophical Papers* (Oxford: Oxford University Press, 2005), 115–30, and Burgess, 'Quinusab omni naevo vindicatus'; the latter is sympathetic to Quine's aims, the former is not.

[8] See R. Carnap, *Meaning and Necessity: A Study in Semantics and Modal Logic* (Chicago: University of Chicago Press, 1947).

prominently in the informal exposition of the natural sciences [RM 158]. However, to capture Quine's intentions clearly the term *analytic* necessity is preferable.

The crucial Lemma of Quine's master-argument for de re modal scepticism, then, is this:

(Lemma) If the quantification and the modality are both taken as ordinarily understood, then there is no obvious sense to be given to de re modal predication. [RM 150]

What Quine means here by quantification 'ordinarily understood' is quantification over what he would call extensional entities: these include the objects of folk theory (people, tables, tigers, mountains, stars . . . ) and the objects, both concrete and abstract, of science (electrons, spacetime points, sets . . . ). What Quine means by modality 'ordinarily understood' is modality as ordinarily understood by those (Carnapian) philosophers he was addressing directly: thus, analytic modality. That there is no obvious sense to be given to de re modal predication so understood is a natural and compelling thought, for what it appears to demand is this.[9] Taking x only as x, it makes perfectly good sense to say that it is analytically necessary of x that it is F. And if having to make good sense of that were bad enough, we must do so while ensuring that we avoid the disaster of committing to cases in which it is analytically necessary of x that it is F and it is not analytically necessary of y that it is F and x=y (cf. (N*1)–(N*3) above).

## 9.3 Strategies of Responses to Quine's Case for Scepticism

In face of Quine's Lemma, my preferred map of the (four) strategic responses available to a 'friend of modality' is as follows:

ACCEPT QUINE'S LEMMA?

/ \

NO (Strategy 1) YES

|

PERSIST WITH THE DEFENCE OF GRADE 3?

/ \

NO (Strategy 2) YES

/ \

Strategy (3)<br>Maintain 'ordinary' modality &<br>invoke 'extraordinary' quantification<br>(Language-dependence)

Strategy (4)<br>Maintain 'ordinary' quantification &<br>invoke 'extraordinary' modality<br>(Language-independence)

[9] Here, and in what follows, it is easier for the reader to think only of cases of atomic 'F'. Many qualifications and complications would have to be introduced in order to take into account at every stage

Strategy (1) is to reject the Lemma. And earning the right to do that would involve undertaking to show either: (a) how Quine is wrong in his view of what doubly ordinary de re modality requires us to make sense of; or (b) how, despite appearances, we can make sense of such a de re combination of quantification ordinarily understood with analytic modality. All other strategies proceed from acceptance of Quine's Lemma.

Strategy (2) is to retreat and defend only 'de dicto' modal commitments: that is, to abandon de-re-modal-predication-cum-quantified-modal-logic and draw the line of defensible commitment under either Grade 1 or under Grade 2 of modal involvement. Pursuit of this strategy requires defence of the classification of some statements as analytic (and others as synthetic) and will still, thereby, involve confrontation with *a* Quinean modal sceptic.[10] But modal scepticism of that kind, and the related confrontation, raises quite different issues from those in prospect here.

Strategies (3) and (4) branch off from the acceptance of Quine's Lemma and involve attempts to defend de re modal predication under that constraint. The attempts in question depart from the 'ordinary' understanding of one of the two elements involved in Quine's conception thus far of de re modal predication. Strategy (3) is to propose an 'extraordinary' understanding of the quantification that is involved in the de re modal predication. Acknowledging those Quine takes to be its proponents, this might be called the Carnap-Church strategy [RM 150–4]: more informatively, it might be called the language-dependence strategy. Strategy (4) is to propose an 'extraordinary' understanding of the modality that is involved in the de re modal predication. Acknowledging him who Quine takes to be its proponent, this might be called the Smullyan strategy:[11] more informatively, it might be called the language-independence strategy.

The aim of the language-dependence strategy, (3), is to make sense of de re modal predication by: (i) reconceiving the domain over which we quantify, and with a view to (ii) making the domain combine safely with the (intended) linguistic—analytic—character of the modality. So the re-conception of the domain of quantification is

the special case where 'F' can be instantiated by a complex predicate that is apt to express a logical truth (e.g. Ax v ~Ax). I claim, but won't attempt to argue here, that the special considerations that apply to those cases do not put them beyond the thrust of Quine's critique. In any case, were de re modal predication to prove defensible in these and only these cases it would be de re predication of a modality that is rather dull and whose population is not obviously well-motivated.

[10] The locus classicus for this is Quine, 'Two Dogmas of Empiricism', *Philosophical Review* 60 (1951): 20–43.

[11] See RM 154ff; TGMI 174; and A. Smullyan, 'Modality and Description', *Journal of Symbolic Logic* 13 (1948): 31–7. It has been pointed out to me (by an anonymous referee, to whom I am grateful) that what I am calling 'the' Smullyan strategy is strictly only *a* Smullyan strategy. For Smullyan's Review of W. V. Quine, 'The Problem of Interpreting modal Logic', *Journal of Symbolic Logic* 12 (1947), p. 140, seems to have also been tempted by the language-dependence strategy (3) on the grounds that co-referential proper names might always be reckoned synonymous.

as a set of entities of a special kind with natures that stand in intimate relations to ways of specifying them. The (historically) salient candidates for being entities of such a special kind are the individual concepts, or senses, that are postulated (in the tradition of Frege, Church, and Carnap) to explain the phenomenon of true but non-analytic identity statements—for example: *Cicero is Tully*, or, *the first Postmaster General is the inventor of bifocals* or *the discoverer of Uranium is Marie Sklodowska-Curie*. Thus, one such entity might be picked out as *the-individual-concept-expressed-by-'Cicero'* and another as *the-individual-concept-expressed-by-'Tully'*. However, puzzlement about the ontology that might be proposed to fit this strange, language-dependent bill, need not detain us. For Quine comes to think that he has a lethal objection to the strategy that depends only on the bill itself [RM 155].

To approach Quine's objection, recall the number-of-planets paradox. The natural diagnosis of the paradox is that it arises precisely because we ordinarily take the individual things that we quantify over, for example the numbers, to be susceptible to analytically inequivalent specification. Thus we have, 'nine', 'the number of planets' but the (presumed) non-analyticity of, 'Nine is the number of planets'. The most direct way of dispatching the paradox, then, is to legislate against such inequivalence: that is, to stipulate that your domain of quantification, in quantified modal logic, will contain only entities that can never be picked out (as variable values) by two conditions that are analytically inequivalent. Bypassing metaphysical worries about the source of such a guarantee, or the conditions of identity of the entities posited,[12] we can proceed immediately to the destination at which Quine eventually arrives in the two classic papers. For Quine's ultimate complaint is that no such legislation can be effective: for it is provable that there are no such special entities of the kind it requires. The suggested proof is as follows. Assume that there is in the language a true but analytically contingent sentence p. (And afford that assumption more security by noting that the predication of analytic necessity of some truths is pointless unless that achieves contrast with others.) Then, given any condition, C, that specifies any x, we can immediately construct a condition, C* that also specifies x but is analytically inequivalent to C—that is: C*(x) =df C(x)& p. So given modal distinctions of the very kind that the proponent of the strategy is trying to defend, there are no things that meet the requirement of being specifiable in a way that is analytically equivalent to no other. And, thus, Quine takes the Church-Carnap strategy to be defeated by a proof. My purposes here require only that we register Quine's view that the failure that besets the language-dependence strategy is of exactly this level: it is demonstrably ineffective. I shall not comment further on Quine's proof, or on how a language-dependence strategy might otherwise be prosecuted.

[12] See Quine, 'Notes on Existence and Necessity', *Journal of Philosophy* 40 (1943): 113–27.

## 9.4 Quine on the Language-independence Strategy

The aim of the language-independence strategy, (4), is to make sense of de re modal predication by: (i) reconceiving the character of the modality involved, and with a view to (ii) making it combine safely with the (presumed) language-independent character of the entities in the domain over which we ordinarily quantify. So the reconception of the modality is as a language-independent, non-analytic but still strict (not-merely-causal) modality, that applies to ordinary things independently of any considerations about how they are specified. Intending to capture exactly that (re)conception, I will call such modality *metaphysical* modality.[13] The final, and most important business of this section will be to register three points about Quine's exposition of this language-independence strategy.

Firstly, Quine's attitude to the Smullyan strategy is quite different from his attitude to the Church-Carnap strategy. For Quine does *not* claim that the prosecution of the Smullyan strategy for expurgating referential opacity is subject to decisive objection. Quine, of course, takes the prosecution of the Smullyan/language-independence strategy to be ill-advised and, perhaps, even misguided. However, this attitude stands in marked contrast with his attitude towards the Church-Carnap/language-dependence strategy, which he takes to be a non-starter for being both demonstrably ineffective and, moreover, ineffective for reasons that even its proponents ought to recognize and admit.

Secondly, Quine predicts that prosecution of the Smullyanite, language-independence strategy will bring three very specific commitments: indeed he clearly and explicitly does so in both *Reference and Modality* (154–6) and in *Three Grades* (175–6). The commitments in question are to:

(Q1) the metaphysical doctrine of Aristotelian essentialism (as Quine represents that);

(Q2) a logic of variables and singular terms that is more complicated, and weaker, than the orthodox approach that is embedded in orthodox classical first-order logic (as Quine understands that);

[13] A caveat is in order concerning the 'ordinariness' of the quantification that I have built into this (the Smullyan) strategy. The initially intended ordinariness involves, naturally, the objects of quantification being *actual* things as opposed to their being *merely possible* things. Yet, it might be required by a full prosecution of the Smullyan strategy across all the formulas of quantified modal logic (as per the semantics of Kripke 'Semantical Considerations on Modal Logic', *Acta Philosophica Fennica* 16 (1963), 83–94) that the required domain of quantification turns out to not be very 'ordinary', in this respect, at all. In the case where a modal operator, especially a possibility operator, has wide scope with respect to a quantifier, as in *it might have been that there were more trees than there actually are*, the values of the variables might be taken either as non-actual-but-merely-possible trees, or as actual things that are not trees but might have been trees. Any such postulates will involve the friend of modality in a further confrontation with *a* Quinean modal sceptic of a kind: in this case, the sceptic about *possibilia* (the locus classicus for this is Quine, 'On What There Is', *Review of Metaphysics* 2 (1948): 21–38). But as remarked in the case of retreating to the defence of only de dicto modality, this move takes us into quite different philosophical territory. The issues I am presently exploring arise even if de re modal predications are restricted by stipulation to those involving quantification over only un-controversially existing actual, and otherwise ordinary, things.

(Q3) the status as a thesis of quantified modal logic, the principle of the necessity of identity as expressed by the formula, (□=):

$$(\Box=)\ \forall x \forall y(x=y \rightarrow \Box x=y).$$ [14]

Thirdly, Quine's dialectic has—of course—particular dialectical opponents in view. Those dialectical opponents were thinkers who shared with Quine allegiance either to a certain conception of logic, or to some form of logical empiricism, but felt able to do so while remaining friends of modality: thus, C. I. Lewis, Church, and Carnap [RM 155–56]. It is this consideration that explains why Quine presumes that the three commitments that he lists will automatically be counted as costs, and as cumulatively unacceptable costs, of implementing the language-independence strategy.

In light of the foregoing account, certain lazy 'takes' on Quine's scepticism about de re modality are exposed as canards—notably: that Quine argues that de re modal predication is absolutely unintelligible; that he is wrong-footed (or even refuted) by the Smullyan response to the number-of-planets paradox; or that he makes an unwarranted presumption that various commitments should be counted as costs. I contend that Quine deserves exoneration from such crass and unjust misrepresentations and that his achievements in this regard deserve greater recognition. Accordingly, I will attempt to show that there are few surprises, and certainly nothing that is apt to move Quine, in the salient philosophical defences of de re modal predication that followed his sceptical critique. My theses are: (a) that that the principals—Kripke, Lewis, and Fine—all undertake non-sceptical defences of de re modal predication that conform to the, Smullyan, language-independence strategy; and (b) none does so in a way that falsifies Quine's prediction of the commitments involved.

## 9.5 Kripke

The positioning of Kripke's defence of de re modal predication in relation to Quine's predictions is a relatively straightforward matter. For Kripke, as far as I am aware, does not dispute Quine's claim about which commitments a defence of de re modal predication will bring. Furthermore, Kripke (1980) consciously embraces and defends all three of the theses to which Quine claims the proponent of the language-independence strategy will be committed.[15]

[14] I use the term 'thesis' to gloss over (otherwise important) distinctions that are not immediately relevant: for example, that between axiom and theorem, and between theorem and valid formula.

[15] To be thorough, one might consider the subtle prospect that Kripke (merely) chooses to embrace all three theses while, in some sense, he is not strictly (de jure) committed to them. Here, I will put aside that subtlety in Kripke's case and in the cases of Lewis and Fine also: except to note one piece of important work that has been done under this heading in Kripke's case. Thus, given certain 'essential sentences' that represent Quine's modal version of Aristotelian essentialism, and a semantic theory of quantified modal logics as per Kripke, 'Semantical Considerations in Modal Logic', Parsons ('Essentialism and Quantified Modal Logic', *Philosophical Review* 78 (1969): 35–52) shows what commitments concerning logical status of such sentences do and do not follow from (commitment to) the Kripkean semantic theory alone.

First, Quine's version of Aristotelian essentialism in *Reference and Modality* is a clearly and explicitly modal version of the doctrine:

> An object, of itself and by whatever name or none, must be seen as having some of its traits necessarily and others contingently, despite the fact that the latter traits follow just as analytically from some ways of specifying the object as the former traits do from other ways of specifying it. [RM 155]

It is Kripke who is primarily responsible for developing and defending the notion of a strict but language-independent conception of modality—the metaphysical modality—that this version of Aristotelian essentialism requires.[16] In that cause, Kripke naturally, and famously, advocates the truth of various cases of de re predication of such a metaphysical modality—for example: that some things, such as Socrates, are necessarily human and some things, such as The Metre Rod, are contingently of the length that they actually are.

Second, it is a characteristic feature of Kripkean semantic theories of (alethic and normal) quantified modal logics, as summarized and discussed in Kripke's 1963 'Semantical Considerations in Modal Logic', that they pronounce as valid (over an appropriate range of model structures) the formula, (□=). In *Naming and Necessity* Kripke also offers various considerations in support of this principle of necessity of identity so construed.

Quine's remaining prediction is that the prosecution of the Smullyan/language-independence strategy for justifying de re modal predication will result in a logic of variables and singular terms that is more complicated, and weaker than, the Russellian approach that is embedded in orthodox classical first-order logic (as Quine understands it). A full examination of this matter, in the case of any of the philosophers to be discussed, calls for the separation of a number of strands in Quine's prediction: for example, restrictions on the introduction and elimination rules of quantifiers in the logic is one thing, the 'construction' of singular terms from term-free resources is another. A full examination also calls for proper consideration of such thorny questions as how the singular terms in the logic might be supposed, by different parties, to relate to the idioms of natural language. However, in this paper I propose to take what I hope are informative shortcuts rather than offer a full tour. In that spirit, then—and thirdly—the following specific theses are all identified by Quine as integral elements of the Smullyan strategy and Kripke endorses all three:

(T1) There are fundamental semantic differences between names and definite descriptions;[17]

However while Parsons' results are non-trivial, they by no means exhaust the sources and kinds of commitment to essentialism that are important in the bigger philosophical picture.

[16] S. Kripke, *Naming and Necessity* (Oxford: Blackwell, 1980).

[17] RM 154; and Kripke, *Naming and Necessity*, pp. 24–6.

(T2) changes in the scope of descriptions with respect to modal operators do make for diffences in truth-value even when the description (actually) refers;[18]

(T3) Intersubstitutivity of variables takes place under restrictive semantic conventions governing the interpretation of variables. That is, the inter-substitutable variables are treated rigidly, in always being assigned to the same object in every world: absent that convention, and other things being equal, (□=) would not be validated.[19]

Thus, I conclude, Kripke's is a perfect example of a Smullyanite defence of de re modal predication as Quine foresees it. With both of the other philosophers to be considered, the evaluation is more complicated. And the common source of complication is their shared conviction that the metaphysical modality involved is non-primitive.[20]

## 9.6 Lewis

On seeing Quine's predictions, it is tempting to leap to the conclusion that Lewis confounds all three. For does not Lewis reject Aristotelian essentialist metaphysics and reject the principle of the necessity of identity and uphold a neo-Russellian descriptivism about singular terms? In each case, there is an important sense in which that is indeed so. Yet, Lewis is not directly at odds with Quine in these matters: and that is because Lewis is not party to a crucial presupposition on which Quine's predictions rely.

Quine, crucially, presupposes throughout his critique of de re modal predication that modal content is represented directly by a modal logical operator, in a non-extensional modal logic: and for Quine, to appeal to such a modal logic is to put modal vocabulary as an element of canonical or primitive notation. For those friends of modality that Quine had in mind when posing his sceptical challenge were all primarily defenders of quantified modal logic. And Lewis's treatment of de re modal predication is free of the commitments that Quine predicts precisely because Lewis's treatment is free of the constraints that come with the commitment to interpret de re modal predication in a special modal logic. That a non-logical, non-canonical, defence of de re modal predication might be mounted is not in conflict with any claim that was earlier made here by me—either for my own part or on behalf of Quine. What was claimed earlier, and claimed to be a matter of impeccable

[18] RM 154 n.9; and Kripke, *Naming and Necessity*, pp. 10–14. So returning to the number-of-planets paradox and reporting the Smullyanite solution, it is held true that *the number of planets* (that very number, 9) is necessarily greater than 7 but false that it is necessary that *the number of planets* is greater than 7 (because there might only have been five planets).

[19] TGMI 175; and Kripke, 'Semantical Considerations in Modal Logic'.

[20] The exact point of contrast is that Lewis and Fine deny that the metaphysical modality is primitive, while Kripke does not deny that. More precisely, Kripke does not take issue with Quine's presumption that a defence of de re modal predication would be a defence of it as 'primitive', in the sense of its being part of logic proper and so of canonical notation. None of this, of course, is to attribute to Kripke the assertion that metaphysical modality is 'primitive' in any respect.

philosophical neutrality, was that the treatment of the inferential properties of Grade 3 constructions is the purpose of quantified modal logic(s). Quite so. *If* we indulge in quantified modal logic then that is why we do it: that is what it is for. But that is not to say that the treatment of the inferential properties of Grade 3 constructions must be handled in that way, and to understand and orientate the Lewisian position on this matter correctly, a sequence of three points must be registered.

The very first thing that Lewis does in approaching the interpretation of de re modal predication,[21] is to distinguish between two ways of doing so: the way of quantified modal logic, and the way of translation into a theory with a first-order, non-modal, logic. Next, Lewis asserts that the former, modal-logical, approach is not inevitable and then he proceeds to demonstrate that by presenting a (counterpart-theoretic) version of the other, first-order non-modal, approach.[22] Thus, the counterpart-theoretic interpretation of discourse involving de re modal predication is one that involves logic only in a perfectly Quinean form: it is fully extensional classical first-order logic. The modal content of de re modal discourse is explicated in a non-logical first-order theory, to which non-modal logic is applied: the domain of the theory in question is a domain of non-actually-restricted individuals (possibilia) and the postulates of the theory govern counterpart relations over those individuals.[23, 24] The statement of these intentions in Lewis's 1968 paper sets out a broad agenda from which he, in his subsequent theorizing of de re modality

[21] David Lewis, 'Counterpart Theory and Quantified Modal Logic', *Journal of Philosophy* 65 (1968): 113–26; page references are to the reprint of the essay in Lewis, *Philosophical Papers* vol. I (Oxford: Oxford University Press, 1983), 26–38, p. 26.

[22] David Lewis, 'Counterpart Theory and Quantified Modal Logic', p. 26.

[23] One potential cause of confusion here is the fact that the notation of quantified modal logic (QML) does have a significant role in the development of counterpart theory (CT) in Lewis's 1968 paper. For the technical part of the paper offers a systematic translation of the formulas of QML into those of CT. But the point of that is to persuade those who are already convinced of the expressive virtues of QML that CT shares these. That is, if you think that de re modal discourse is represented systematically, and with a certain degree of expressive completeness, by the formulas of QML, you must accept that also for the formulas of CT. The non-technical part of the paper is to (begin to) argue that the CT representation of de re modal discourse is (at least) as adequate as the QML representation in broader semantic and philosophical respects. One can see how Lewis's approach might be adapted and re-organized so that it would become, through counterpart-theoretic models, a genuine interpretation of quantified modal logic. See G. Hughes & M. Cresswell, *A New Introduction to Modal Logic* (Routledge: London, 1996), pp. 353ff. But that was not Lewis's project, and it is crucial in understanding the relationship of Lewis to Quine that this is understood.

[24] Another potential cause of confusion here is the fact that Lewis allows the way that we choose to specify things a role in picking out which counterparts enter into the truth-conditions of de re modal sentences in which those specifications feature. For is that not a clear renunciation of the Smullyan strategy that proceeds from language-independent modality? No. For we don't conceive the modal facts as language-dependent just because we rely on considerations about language to pick out which language-independent facts (facts of objective similarity, for Lewis) they reduce to. For extensive discussion of that point see Divers, 'Quinean Scepticism about De Re Modality after David Lewis', *European Journal of Philosophy* 15 (2007): 40–62. So Lewis's position as a Smullyanite defender of de re modal predication is secure.

never departs.[25] And to take proper account of this agenda we must handle matters with some care.

First, Lewis claims that what is represented here by (□=) is a modal principle (of necessity of identity) that should not be accepted.[26] It is crucial to take seriously Lewis's choice of terminology. What Lewis does not claim is that (□=) is invalid or that it is a non-theorem. For 'invalid' and 'non-theorem' (and their cognates) are terms that are apt in discussing the model-theoretic and proof-theoretic status of formulas of your logic, and Lewis is not considering the formulas of QML as formulas of his logic. In light of this crucial distinction (between the canonical and the non-canonical) Quine's intended and frequent observation in the matter might be put thus. If you locate de re modal predication in your logic—and, a fortiori, in your canonical notation—and you also refuse (□=), then you make a nonsense of the logic of identity. For then you are denying that a primitive or canonical predication (such as $\Box F$) is equally predicable of primitively or canonically specified identicals (by hypothesis, $x$ and $y$). Certainly, if you take an apparent failure of intersubstitutivity of identicals to be merely a quirk of representation at a non-canonical level—like English, or a logic into which non-variable singular terms have been introduced—you always have the option of mitigating that by showing that it disappears at the canonical level. Indeed, Quine [TGMI 173–5] makes precisely this point on behalf of the friends of modality. So Quine's claim about (□=) is about what must be involved in defending de re modal predication as a feature of canonical notation, beyond which there is nowhere to run and nowhere to hide. And the Lewisian rejection of the principle (□=) does not provide a counterexample to Quine, for Lewis does not reject (□=) while attempting to treat de re modal predication as a feature of a canonical notation.

Second, the treatment of singular terms and variables in Lewis, as in the other cases, must be somewhat rough and ready here. However, here are the salient points. Given Lewis's thoroughly Quinean conception of logic as classical first-order logic with identity, and adding to that the primitive predicates of counterpart theory, there is no obvious pressure from the logic of identity to depart from the Russellian conventions for introducing and eliminating singular terms via the description operator. There is no obvious need, thus far, to weaken or complicate the logic of identity or introduce supporting lemmas.[27] However, if we ascend to a level at which

[25] Especially not so in the further classic sources of Lewis on modality: *Counterfactuals* (Oxford: Blackwell, 1973) and *On the Plurality of Worlds* (Oxford: Blackwell, 1986).

[26] David Lewis, 'Counterpart Theory and Quantified Modal Logic', p. 36. To be truly precise, Lewis's claim concerns the open analogue of (□=). But that is irrelevant for present purposes.

[27] To be fair, there are non-obvious objections to the effect that Lewis's CT translations do require him to mess with the handling of variables in first-order logic. The objections are due to Hazen, 'Counterpart-Theoretic Semantics for Modal Logic', *Journal of Philosophy* 76 (1979): 319–38, and Kripke (*Naming and Necessity*, p. 45, n. 13) and here I can address them only insofar as I am prepared to state my conviction that Lewis deals adequately with these in his 'Postscript to Counterpart Theory and Quantified Modal Logic', p. 45.

modal expressions (the modal operators) are also in play—introduced, as it were, via their counterpart-theoretic 'translations'—then Lewis is a self-identifying Smullyanite[28] and he embraces the relevant Quinean predictions. That is, in the first place, it is acknowledged that de re predications involving modal expressions are in general scope ambiguous: they are translatable into the 'canonical' notation of counterpart theory in scope-sensitive ways that are not logically equivalent. And the effect of this is precisely to force the need for 'supporting lemmas' in governing whether descriptively analysed terms are available as variable substituends for universal elimination or existential introduction. For such terms will be available only when the wide/de re translation is equivalent to the narrow/de dicto translation, and that is when certain further lemmas of counterpart theory hold.[29]

Finally, Lewis is explicit in his endorsement of (his finding 'congenial') Quine's version of Aristotelian essentialism.[30] So, inter alia, what Lewis finds congenial is that objects have in themselves, and independently of any consideration of language, some traits necessarily and other traits contingently. That is a modal commitment. But such a modal commitment need not be in itself much by way of a metaphysical commitment and not, a fortiori, a commitment to specifically Aristotelian metaphysics in any demanding sense. We appreciate the need to distinguish one kind, or level, of commitment from the other once, as we have established, the distinction between the canonical and the non-canonical (or primitive and non-primitive) is in play. For, as we have seen already in the discussion of (□=), what one is committed to is one thing, and what one is committed to there being *at the canonical or primitive level* is another. Quine might be charged with making a mistake in taking fairly minimal commitments about what is necessarily (or essentially) this way or that and then *characterizing these as* acceptance of a very non-minimal metaphysical doctrine—the metaphysical doctrine of Aristotelian essentialism (see, e.g., 'Three Grades of Modal Involvement', p. 176). In charity, though, we ought to recall Quine's presumption that the discussion is about what is defensible by way of a (modal) logic and, so, at the canonical level. And we ought to remind ourselves that what is perhaps the most famous Quinean dictum in this matter is precisely and explicitly about the consequences of championing a logic—thus: 'Aristotelian essentialism should be every bit as congenial to (the champion of quantified modal logic) as quantified modal logic itself'.[31]

In sum, then, Lewis's position affords no counterexample to anything that Quine predicted in this regard. Lewis does not commit to a deeply metaphysical Aristotelian Essentialism: that 'fundamental' reality is such that things have such modal features.

[28] Lewis, 'Counterpart Theory and Quantified Modal Logic', p. 33, n.16.
[29] Lewis, 'Counterpart Theory and Quantified Modal Logic', pp. 33–4.
[30] Lewis, 'Counterpart Theory and Quantified Modal Logic', p. 32.
[31] Quine, 'Reply to Professor Marcus', *Synthese* 20 (1961): 177–84, p. 184.

But nor does Lewis commit to treating de re modal predication as a feature of 'fundamental' notation. And Quine's prediction, insofar as it might be put in these terms, was only that commitment to the latter would require commitment to the former.

## 9.7 Fine

There is no respect in which Fine needs to depart, or in which he actually does depart, from the commitments, (Q1)–(Q3), that Quine predicts. For, as I understand it, Fine's, non-sceptical, Smullyanite, defence of de re (metaphysical) modal predication replicates the Kripkean position in acceptance of the modal doctrine of Aristotelian Essentialism, the necessity of identity in the form of (□=) and the non-Russellian treatment of singular terms and variables.[32] This is the point that is central to present purposes and it ought to be registered and acknowledged. But there is more here that needs to be explained and more to appreciate about Quine in the process.

What remains to be understood about the agreement between Kripke and Fine, and what makes Fine's work among the most significant on the topic, is this. After metaphysical modality is up-and-running, as it were, Fine may be viewed—in all immediately relevant respects, and applying a broad brush—as being as Kripkean as Kripke. And that is why, as noted, Fine, like Kripke, is in the position of fulfilling all of Quine's predictions about the language-independence defence of de re modal predication. Where Fine proceeds beyond Kripke is in taking us into a metaphysical 'jungle' of Aristotelian Essentialism that Quine would have found deeper and darker than any jungle that he himself had dared to envisage. Kripke embraces the predicted commitments, (Q1)–(Q3), without indicating any departure from the Quinean presupposition that the defence of de re modal predication is to be a defence of de re modal predication as primitive. However, in 'Essence and Modality' Fine rejects the primitive status of de re modal predication and regards Kripkean modal ideology (including all of the Quine-predicted commitments) as metaphysically and logically supported by deeper essentialist commitments—this position presupposing, of course, that the esentialist commitments are not simply equivalent to de re (metaphysically) modal commitments. In this regard, it is informative to consider Quine's second version of the metaphysical doctrine of Aristotelian essentialism given in *Three Grades*—thus:

> This is the doctrine that some of the attributes of a thing (quite independently of the language in which the thing is referred to, if at all) may be essential to the thing and others accidental. E.g. a man, or talking animal, or featherless biped (for they are all in fact the same things) is essentially rational and accidentally two-legged and talkative not merely qua man but qua itself. [TGMI 176]

[32] The (now) classic source that I have in mind and to which I shall refer most frequently is K. Fine, 'Essence and Modality', *Philosophical Perspectives* 8 (1994): 1–16. But see also, for example, Fine, 'The Logic of Essence', *The Journal of Philosophical Logic* 24 (1995): 241–73, and his 'Introduction' to *Modality and Tense: Philosophical Papers* (Oxford: Oxford University Press, 2005), 1–18.

The first version of the doctrine, given by earlier quotation from *Reference and Modality*, I characterized as modal, because it involves explicit predications of the paradigmatically modal modifiers, 'necessarily' and 'contingently'. This second version of the doctrine puts in the place previously occupied by those modal modifiers the explicitly essentialist modifiers, 'essentially' and 'accidentally'. Before Fine, most—and notably both Kripke and Lewis—acquiesced in the presumption, apparently shared by Quine, that nothing (much) was at stake in distinguishing the modal version of essentialism from the essentialist version of essentialism: hence the free and unconcerned movement back and forth between modal and essentialist predications.[33] But the contention of Fine is precisely that modal predication—and especially de re metaphysically modal predication—is to be explained in terms of different and more fundamental considerations about essence. Thus, it is true of Socrates that he is (metaphysically) necessarily human: but the truth of this de re modal predication obtains in virtue of (the modal fact is 'grounded in') non-equivalent facts about the essence of Socrates. And the facts about essence are, in turn and ultimately, the identity-making facts (what makes it the case that Socrates is the very thing that he is) as contrasted with the attribute-fixing facts (what makes it the case that Socrates—identity fixed—is how he is). Fine's 1994 asymmetry thesis is that x's being (metaphysically) necessarily F is necessary but not (even materially) sufficient for x's being essentially F. Some of Fine's illustrations of insufficiency exploit exactly the same trick that Quine exploited in his anti-Church-Carnap proof above. The trick is to invoke that expansive conception of non-modal predicates that includes open sentences that have closed sentences as parts. Thus in Fine's defence of insufficiency we have the likes of *necessarily x is a philosopher or 2+2=4* being true of Socrates but *essentially x is a philosopher or 2+2=4* being (we are told) not true of Socrates. Various other kinds of illustrations are also invoked to exploit the 'intuition' that only some of what is metaphysically necessary of Socrates is attributable to his being the very thing that he is. For example, it is metaphysically necessary of Socrates that he is a member of the set {Socrates} and it is metaphysically necessary of {Socrates} that it has Socrates as a member. But while having Socrates as a member is essential to (grounded in the identity of) {Socrates}, we are told that being a member of {Socrates} is not essential to—because not grounded in the identity of—Socrates.

Thus, Fine exceeds Quine's prescience only by proceeding further in a direction along which Quine predicted only the minimum distance of travel that would be involved. From Quine's standpoint, the natural reaction to Fine's defence of de re modal predication is as follows. It was never ruled out, and now it is proved, that the pursuit of a coherent worldview in which to embed de re modal predication might

[33] This is not to say that everyone was happy to accept that all essentialist claims were adequately expressed through the notation of standard first-order quantified modal logic. See, for example, David Wiggins, 'The *De Re "Must"*: A Note on the Logical Form of Essentialist Claims', in *Truth and Meaning: Essays in Semantics*, edited by G. Evans & J. McDowell (Oxford: Oxford University Press, 1976), 285–312.

lead to even deeper metaphysical commitments than were predicted as the minimum in that regard. It goes with the genuinely modal territory, beyond Grade 1, and all agree, that we must abandon a purely extensional conception of logic. In modal logic, we can no longer freely intersubstitute expressions with the same extensions—for example predicates that apply to the same things, or statements with the same truth-value. For it is true that *2+2=4* and true that *Quine is a philosopher*. But when we substitute the latter for the former in the truth that *it is necesssary that 2+2=4* we get the falsehood *it is necessary that Quine is a philosopher*. Quine [TGMI 168ff] thought also—for reasons I have not had time to go into—that the only coherent position for a friend of modality was one in which there was no stopping at the minimal modal logic of Grade 2. So acceptance of modality leads to acceptance of quantified modal logic, and with that evident intensionalist metaphysical commitments. For one would accept (in some cases) that it is a fundamental feature of the world, because it is so according to the canonical statement of best theory, that *x is necessarily F*. But if Fine is right, what is in prospect is a further push from Grade 3 on to a further (fourth) grade of non-extensional involvement that is not merely modal and not merely intensional. For in essential contexts, we can no longer freely intersubstitute expressions even when they are modally (that is, intensionally) equivalent—for example predicates that apply to exactly the same things in every possible circumstance or statements that necessarily have the same truth-value. For, on Fine's view that will lead us from truth to falsehood in some cases where we substitute even modally equivalent predicates into *x is essentially* F. For it is true that *Socrates is necessarily such that he is identical to Socrates* and true that *Socrates is necessarily such that either he or 2 is an even number*, but while it is also true that *Socrates is essentially such that he is identical to Socrates* it is (we are told) false that Socrates is *essentially such that either he or 2 is an even number*. Thus, in Fine's scheme of things, acceptance of modality leads ultimately to not only intensionalist metaphysical commitments but to hyperintensionalist metaphysical commitments. For one would accept (in some cases) that it is a fundamental feature of the world, because it is so according to the canonical statement of best theory, that *x is essentially F*. So for Quine, the prospect raised by Fine's prosecution of the Smullyan strategy is that of inflating the costs of defending de re modal predication to a greater level, by a whole order of magnitude, above those that he had dared C. I. Lewis, Church, and Carnap to contemplate.

## 9.8 Conclusion

What Quine foresaw was this. (1) The project of defending quantified modal logic, and thereby de re modal predication as an element of canonical notation, is not doomed to be ineffective. But, (2) the only defensive strategy that has a chance of proving effective is the Smullyan strategy on which we reconceive strict modality as a language-independent modality. And, (3) once that step is taken, specific commitments to (at least) the following, are bound to ensue: the modal version of the doctrine of Aristotelian essentialism, some significant departures from the classical (Russellian, description-centred) treatment of variables and singular terms, and a quantified modal logic that has as a thesis the necessity of identity, (□=). What Kripke does subsequently is to embrace entirely Quine's theses, (Q1)–(Q3) and try to make the package an attractive one. In doing so, Kripke accepts, with the first move, a presupposition that Quine shared: namely, that a defence of de re modal predication will locate it in the logic proper, and hence as a feature of canonical notation. What Lewis and Fine both do is to refuse that presupposition. Having done so, neither Lewis nor Fine takes issue with, nor confounds, Quine's predictions, (Q2) and (Q3), about the consequences of accepting that presupposition. I doubt that Quine was surprised by the emergence of the, obvious and natural, strategic thought that de re modal predication might be defended as a non-canonical (reducible) aspect of 'total theory'. Perhaps Quine's only surprise would have been at the metaphysical lengths to which these philosophers have been prepared to go to in mounting such a defence. For the metaphysical postulates in question are, with Lewis, an infinity of universes across which every metaphysical possibility is realized and, with Fine, a universe that is more radically non-extensional than Quine took even the ancient worldview of Aristotelian essentialism to suggest.[34]

## Bibliography


Burgess, J., 'Quinusab omni naevo vindicatus' in *Mathematics, Models and Modality* (Cambridge: Cambridge University Press, 2008): 203–35.

Carnap, R., *Meaning and Necessity: A Study in Semantics and Modal Logic* (Chicago: University of Chicago Press, 1947).

Divers, J., 'Quinean Scepticism about De Re Modality after David Lewis', *European Journal of Philosophy* 15 (2007): 40–62.

Fine, K., 'Quine on Quantifying In' in *Proceedings of the Conference on Propositional Attitudes*, edited by C. A. Anderson & J. Owens (Stanford CSLI, 1990, 1–26). Reprinted in Fine, *Modality and Tense: Philosophical Papers* (Oxford: Oxford University Press, 2005), 115–30.

Fine, K., 'Essence and Modality', *Philosophical Perspectives* 8 (1994): 1–16.

Fine, K., 'The Logic of Essence', *The Journal of Philosophical Logic* 24 (1995): 241–73.

Fine, K., *Modality and Tense: Philosophical Papers* (Oxford: Oxford University Press, 2005).

Hazen, A., 'Counterpart-Theoretic Semantics for Modal Logic', *Journal of Philosophy* 76 (1979), 319–38.

Hughes, G. & Cresswell, M., *A New Introduction to Modal Logic* (Routledge: London, 1996).

Kripke, S., 'Semantical Considerations on Modal Logic', *Acta Philosophica Fennica* 16 (1963), 83–94.

Kripke, S., *Naming and Necessity* (Oxford: Blackwell, 1980).

Lewis, D., 'Counterpart Theory and Quantified Modal Logic', *Journal of Philosophy* 65 (1968): 113–26.


[34] I thank Joseph Melia for his comments on an earlier draft of this paper.

Lewis, D., *Counterfactuals* (Oxford: Blackwell, 1973).

Lewis, D., Postscript to 'Counterpart Theory and Quantified Modal Logic' in *Philosophical Papers*, Vol. I (Oxford: Oxford University Press, 1983), 39–46.

Lewis, D., *On The Plurality of Worlds* (Oxford: Blackwell, 1986).

Parsons, T., 'Essentialism and Quantified Modal Logic', *Philosophical Review* 78 (1969), 35–52.

Quine, W. V. O., 'Notes on Existence and Necessity', *Journal of Philosophy* 40 (1943): 113–27.

Quine, W.V.O., 'On What There Is', *Review of Metaphysics* 2 (1948), 21–38.

Quine, W.V.O., 'Two Dogmas of Empiricism', *Philosophical Review* 60 (1951), 20–43.

Quine, W.V.O., 'Three Grades of Modal Involvement', *Proceedings of the XIth International Congress of Philosophy*, Vol. 14 (Amsterdam: North-Holland Publishing Co., 1953). Reprinted in *The Ways of Paradox and Other Essays*, revised edition (Cambridge, MA: Harvard University Press, 1976), 158–76.

Quine, W.V.O., 'Reference and Modality', in *From a Logical Point of View*, 2nd edition (New York: Harper and Row, 1961), 139–59.

Quine, W.V.O., 'Reply to Professor Marcus', *Synthese* 20 (1961): 177–84.

Shieh, S., 'Modality' in *The Oxford Handbook of the History of Analytic Philosophy*, edited by M. Beaney (Oxford: Oxford University Press, 2013), 1043–81.

Smullyan, A., Review of W.V. Quine, 'The Problem of Interpreting Modal Logic', *Journal of Symbolic Logic* 12 (1947): 139–41.

Smullyan, A., 'Modality and Description', *Journal of Symbolic Logic* 13 (1948): 31–7.

Wiggins, D., 'The *De Re "Must"*: A Note on the Logical Form of Essentialist Claims', in *Truth and Meaning: Essays in Semantics*, edited by G. Evans & J. McDowell (Oxford: Oxford University Press, 1976), 285–312.

# Index

The manufacturer's authorised representative in the EU for product safety is Oxford University Press España S.A. of el Parque Empresarial San Fernando de Henares, Avenida de Castilla, 2 – 28830 Madrid (www.oup.es/en or product.safety@oup.com). OUP España S.A. also acts as importer into Spain of products made by the manufacturer.

www.ingramcontent.com/pod-product-compliance
Lightning Source LLC
LaVergne TN
LVHW011819200926
848484LV00005B/1245

* 9 7 8 0 1 9 8 7 8 6 4 3 6 *